I0090019

Sanitation in Humanitarian Settings

Praise for this book

'*Sanitation in Humanitarian Settings* by Robert Reed makes an important contribution to sector knowledge on the topic. Where other guidance focuses primarily on containment, this guide covers the entire sanitation chain and addresses faecal sludge management in detail from both technical and non-technical perspectives. It also provides practical guidance on the related, but too often neglected, issues of sullage and solid waste management, making this an invaluable addition to the humanitarian worker's toolkit.'

Dr. Peter Harvey, United Nations Children's Fund (UNICEF)

'*Sanitation in Humanitarian Settings* captures in one place the recent thinking in emergency sanitation. While there is a focus on emergency excreta disposal and treatment, much of the appeal of this book is the attention given to the other aspects of sanitation such as faecal sludge management in disease outbreaks, grey water disposal, solid waste and recycling. It is a must read for any practitioners in emergency sanitation.'

Andy Bastable, Head of Water and Sanitation, Oxfam Global Humanitarian Team

'In humanitarian contexts the importance and life-saving nature of effective sanitation and faecal sludge management cannot be overstated. This long-awaited book combines decades of practical experiences from various contexts with the latest global sector knowledge and provides structured and detailed guidance on the most crucial technical and non-technical aspects of a sanitation response. I highly recommend it to be used as a constant companion for all those – voluntarily or involuntarily – confronted with planning and implementing humanitarian sanitation interventions.'

Rob Gensch, German Toilet Organization

'In the face of unforeseen circumstances, having access to a reliable and comprehensive guide to emergency sanitation is crucial for ensuring the health and safety of individuals and communities. *Sanitation in Humanitarian Settings* is an invaluable resource that provides readers with a clear understanding of essential full sanitation practices during emergencies. The author's expertise shines through in the ability to distil complex concepts into easy-to-follow guidance, making this book accessible to a wide audience in every aspect of sanitation in emergency situations. Whether navigating natural disasters or humanitarian crises, this emergency sanitation book is indispensable for anyone seeking to maintain a healthy and hygienic environment. Its comprehensive coverage, practical approach, and user-friendly format make it an essential addition to any emergency preparedness kit.'

Debora Bonucci, WASH Adviser at the British Red Cross

Sanitation in Humanitarian Settings

Robert Reed

Practical
ACTION
PUBLISHING

To Janet

Without your support and loving patience this book
would not have been possible

Practical Action Publishing Ltd
25 Albert Street, Rugby,
Warwickshire, CV21 2SD, UK
www.practicalactionpublishing.org

© Robert Reed, 2024

The moral right of the author to be identified as author of the work has been asserted under sections 77 and 78 of the Copyright Design and Patents Act 1988.

All rights reserved. No part of this publication may be reprinted or reproduced or utilized in any form or by any electronic, mechanical, or other means, now known or hereafter invented, including photocopying and recording, or in any information storage or retrieval system, without the written permission of the publishers.

All figures with the exception of those referenced were produced by Rod Shaw and can be found at this address (** WEDC. Loughborough University, 2024) where they may be accessed under a creative commons license for non-commercial use.

Product or corporate names may be trademarks or registered trademarks, and are used only for identification and explanation without intent to infringe.

A catalogue record for this book is available from the British Library.
A catalogue record for this book has been requested from the Library of Congress.

ISBN 978-1-78853-265-5 Paperback
ISBN 978-1-78853-266-2 Hardback
ISBN 978-1-78853-267-9 Electronic book

Citation: Reed, R., (2024) *Sanitation in Humanitarian Settings*, Rugby, UK: Practical Action Publishing http://doi.org/10.3362/9781788532679.

Since 1974, Practical Action Publishing has published and disseminated books and information in support of international development work throughout the world.

Practical Action Publishing is a trading name of Practical Action Publishing Ltd (Company Reg. No. 1159018), the wholly owned publishing company of Practical Action. Practical Action Publishing trades only in support of its parent charity objectives and any profits are covenanted back to Practical Action (Charity Reg. No. 247257, Group VAT Registration No. 880 9924 76).

The views and opinions in this publication are those of the author and do not represent those of Practical Action Publishing Ltd or its parent charity Practical Action.

Reasonable efforts have been made to publish reliable data and information, but the author and publisher cannot assume responsibility for the validity of all materials or for the consequences of their use.

Cover photo shows a team of Bangladeshi workers emptying faecal sludge from an emergency toilet and pumping it to storage tanks from where it is piped to a temporary treatment plant.

Cover photo credit: Robert Reed
Cover design by Rod Shaw, WEDC and Katarzyna Markowska, Practical Action Publishing
Typeset by vPrompt eServices, India

Contents

About the author

Robert (Bob) Reed has been involved in water supply and sanitation for low-income communities and emergencies for 40 years. He has worked in many countries, mainly in Africa and Asia, but for over 30 years he was based at Loughborough University working in a group called WEDC where he undertook a range of duties including lecturing, consultancy, and research. He worked in his first emergency in 1985, building communal toilets for displaced people, and his last involvement was in 2019, constructing an emergency faecal sludge treatment plant. He has many publications to his name covering both the theory and the practice of emergency water supply and sanitation.

About the illustrator

Rod Shaw has over thirty five years of experience of design and illustration for international development, working for WEDC at Loughborough University and for external organizations including The World Health Organization (WHO) and the International Federation of the Red Cross and Red Crescent Societies (IFRC). He is the author and illustrator of the resource book Drawing Water, and a collection of over 600 line illustrations related to water, sanitation, and health in low- and middle-income countries.

Acknowledgements

Much of the information gathered for this book was obtained while I worked at WEDC, Loughborough University. I would like to thank all the staff of that organization, both past and present, for their support, advice, and guidance during my time there. Without them I would never have been able to gain the knowledge and experience that I have.

In particular, I would like to thank Rod Shaw (r.j.shaw@lboro.ac.uk) and Ken Chatterton (ken@hallchat.co.uk) for all the images they have produced over the years, many of which are now being used in this book. I strongly believe that images demonstrate far more than words ever can, especially in engineering. The value of this book would be considerably reduced without those produced by Rod and Ken.

The author would like to express his gratitude to those who agreed to review the book through the many iterations of its development. Without their detailed and extensive comments and suggestions this book would have been of much lesser worth. In particular, I would like to thank Andy Bastable, Debora Bonucci, Jan Davis, Ben Harvey, Peter Harvey, Emmett Kearney, Jenny Lamb, Maryanne Leblanc, Reymond Phillippe, and David Thomas.

List of figures, tables, and boxes

Figures

Tables

Boxes

CHAPTER 1
Introduction

In an emergency, the collection, treatment, and disposal/reuse of human waste are critical to the health, well-being, and environment of affected populations. Addressing this is difficult and complex and requires an understanding of the technical, social, institutional, political, and environmental issues affecting the situation. In large emergencies where international relief agencies are involved, there will usually be specialists available to advise and assist inexperienced field staff in delivering essential services. However, most emergencies are not large. They are local in nature and rely on local officials to provide an emergency response, especially in the early stages of the emergency. It is very likely that these officials will have minimal knowledge and experience of emergency response; indeed, they have probably never been faced with such an event before.

This book sets out to assist such individuals. Working from a technical perspective, it provides an understanding of the critical elements controlling the delivery of human waste-related services together with strategies for decision making and guidance on how to design, construct, and operate key infrastructure.

The book concentrates on the acute response and stabilization phases of an emergency (Box 1.1), but much of the content is equally relevant to the long-term and rehabilitation phases.

1.1 Who should read the book

The book is primarily aimed at junior engineers and technicians who are working or would like to work in emergency human waste management (including excreta, sullage/grey water, and domestic solid waste). The book assumes a working knowledge of basic civil engineering practice such as surveying and simple construction techniques but no or minimal experience of working in emergencies or specifically in human waste management. The book will also be of use to teachers and researchers wishing to extend their knowledge of the subject and/or teach it to others.

1.2 Scope

The book focuses on the three main areas of emergency human waste management: human excreta management, sullage/grey water disposal, and solid waste management. Emergency excreta management is by far the largest component of the book. Not only is it the biggest issue in human

---

Box 1.1 Emergency phases

The response to an emergency can be broadly divided into three phases. The duration of each phase is very variable and depends on local circumstances, but a general understanding is useful for decision making and planning.

Acute response phase: This is implemented immediately following the outbreak of the emergency. It usually covers the first hours and days up to the first few weeks, where effective short-term measures are applied to alleviate the emergency situation quickly until more permanent solutions can be found. The purpose of interventions in the acute response phase is to ensure the survival of the affected population. Essential faecal management-related services needed at this stage include establishing instant and safe excreta management options (particularly excreta containment measures) as they are critical determinants for survival in the initial stages of a disaster. Ensuring a safe environment and avoiding contamination of water sources are also critical. If applicable, this phase may also include the quick rehabilitation of existing infrastructure, the establishment of appropriate drainage solutions, and the provision of tools and equipment to ensure basic operation and maintenance services.

Stabilization phase: This usually starts after the first weeks of an emergency and can last several months to half a year or longer. The main faecal management focus, apart from increasing coverage, is the incremental upgrade and improvement of the temporary emergency structures that would have been installed during the acute response phase, or the replacement of temporary technologies with more robust longer-term solutions. This phase includes the establishment of community-supported structures with a stronger focus on the entire faecal management service chain. This phase often sees a shift from communal services to household-level solutions. Particular emphasis should be given to socio-cultural aspects such as potentially sensitive issues regarding toilets (including use, operation, and maintenance), menstrual hygiene management, and vulnerability to sexual and other forms of violence, as well as hygiene-related issues that imply certain levels of behaviour change. The equitable participation of women and men, children, and marginalized and vulnerable groups in planning, decision making, and local management is key to ensuring that the entire affected population has safe and adequate access to appropriate human waste management services.

Recovery phase: Sometimes referred to as the rehabilitation phase, this usually starts after or even during relief interventions and aims to recreate or improve on the pre-emergency situation of the affected population. It can be seen as a continuation of already executed relief efforts and can prepare the ground for subsequent longer-term interventions and the gradual handing over to medium/long-term partners. Depending on local needs, the general timeframe for recovery and rehabilitation interventions is usually between six months and three years, or, in difficult situations, up to five years.

Source: Adapted from (** Gensch, et al., 2018)

---

waste management; it is also the most complex and challenging. Each chapter focuses on an element of service delivery, from an explanation of the working environment through situation assessment, technology selection, detailed design, and construction to operation and management.

1.3   Referencing

In recent years there has been a great increase in literature related to emergency human waste management, particularly emergency excreta management; this has been reflected in the production of a large range of helpful information.

Unfortunately, this is available in multiple formats, from web pages to books and professional journals, as reflected in the extensive list of references given at the end of this book. As such, much of the information provided here can be found in other publications, which are referenced wherever possible. There are, however, some key publications that have been used extensively and we recommend that the reader consults them directly for more detailed information. They are highlighted in references by the addition of '**' before the author's name. The book also draws on the practical experience of many field practitioners, together with the author's 40 years of working in the sector. Such a wide range and variety of sources makes it very difficult for inexperienced practitioners to obtain a holistic overview of current knowledge and best practice, something this book hopes to offer.

1.4 How to use the book

Most readers will use this book as a reference document, dipping into it according to their needs. It has therefore been provided with a comprehensive table of contents and index plus copious cross-referencing to help users find the most relevant information. The large list of references will also help users find more information on areas of specific interest.

In its current form the book is laid out to be read as a conventional book, although a direct electronic version will be available. It is hoped that a fully electronic version of the book will be available shortly that will include hot links between related sections.

1.5 Terminology

Terminology in emergency relief is confusing and vague, with different organizations using different words for the same thing. This book has tried to standardize its terminology throughout. A discussion of the most important terms used in the book is included below, but a fuller list can be found in Appendix 8.

The toilet
This term causes lots of confusion, including what it actually means, what is and isn't included, and whether it's the correct term anyway! For the purposes of this book, a toilet is defined as being the space, in total, in which users can defecate and urinate conveniently in safety and privacy (Figure 1.1).

It includes the cubicle building, floor, roof, handwashing facilities, and interface for defecation. The room may also include other facilities such as washing, bathing, and laundry facilities. The space may be designed for an individual family, several families, a specific community, or public use.

Affected population
This refers to those whose lives have been dramatically affected by man-made or natural disasters, including people who have been forced to leave their

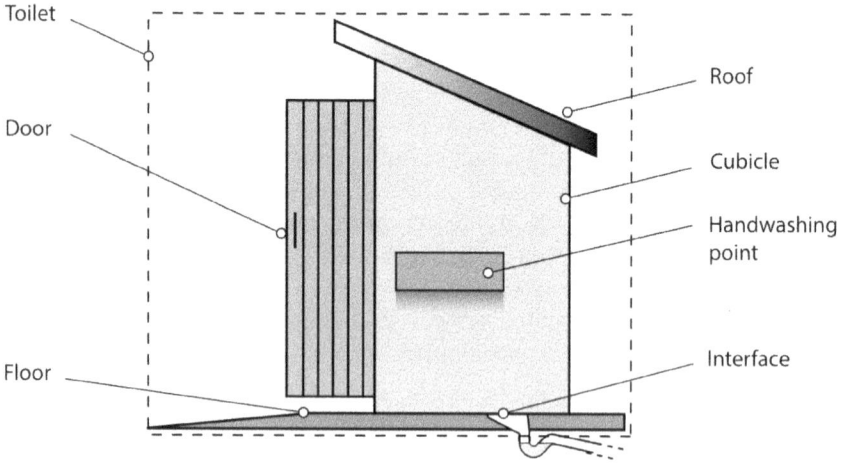

Figure 1.1 Toilet elements

homes and live with other families or move to temporary settlements. It also includes those who have had their home and local environment severely damaged but choose not to move. The affected population may be rural, urban, or anything in between.

Emergency
There's also a lot of confusion between the words 'emergency' and 'disaster' and no fully accepted definitions. According to the WHO (Panafrican Emergency Training Centre, 2002) an emergency is 'a state in which normal procedures are suspended and extraordinary measures are taken in order to avert a disaster', whereas a disaster is 'an occurrence disrupting the normal conditions of existence and causing a level of suffering that exceeds the capacity of adjustment of the affected community'.

In practice, the words are often used interchangeably. For the purposes of this book, the term 'emergency' will be taken to mean any event that disrupts normal conditions where the local community needs external assistance to deal with it.

Emergency human waste management
This is primarily the collection, transport, treatment, and use/disposal of human excreta, but it also includes liquid and solid wastes that could have come into contact with human excreta (e.g. toilet paper, domestic solid waste, menstrual hygiene materials, and sullage). The process of excreta management is shown in Figure 1.2.

Sanitation
Sanitation is a word frequently used when discussing excreta-related management, but it creates a lot of confusion. There is no agreed definition of

Collection ⟶ Transportation ⟶ Treatment ⟶ Disposal / Reuse

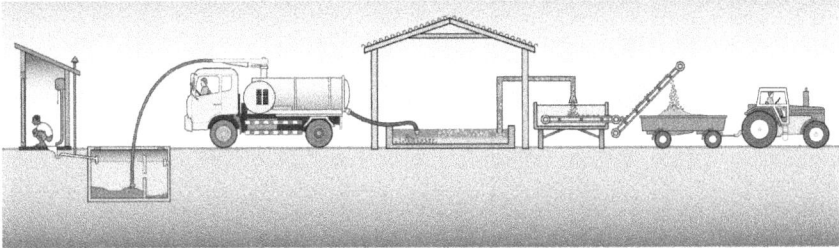

Figure 1.2 Faecal sludge management chain

the word, and this has led to individuals and organizations using it in a variety of ways. For this reason, this book uses the word as little as possible.

1.6   Errors and omissions

This book contains a lot of information and data. The contents have been checked many times, but, in a field where experience and innovation are growing rapidly, it is possible that you will read things here that are incorrect or just out of date. The author would appreciate it if you would notify him of any errors or omissions that you find by writing to the publishers. The changes can then be included in future editions.

CHAPTER 2
Priorities and objectives

Any organization, agency, or individual must understand their reasons for intervening and what they are trying to achieve. Individual organizations often have their own reasons and objectives, but the following describes the international-level priorities, which are based on minimizing public health risks from the outset.

2.1 An environment free from human excreta

Such an environment is essential to life, health, and dignity. Safe human waste management (HWM), especially excreta management, is among the primary interventions during an emergency response to reduce the risk of disease transmission; it is essential for life with dignity and should be addressed with the same speed and effort as the provision of safe water supply. Containment of human waste away from people stops excreta-related disease by reducing direct and indirect routes of disease transmission, as shown in Figure 2.1. Measures should be taken with all HWM interventions to minimize all risks to public health and the environment.

2.2 Community engagement

Community engagement is critically important. It is a continuous process connecting the community to the organization delivering the service to ensure that affected populations have a voice and control over the response and its impact on them. From the outset, community engagement creates an essential understanding of the community's perceptions, needs, coping mechanisms, preferences, existing norms, leadership structures, and priorities, as well as the appropriate actions to take.

The service delivery organization must consult, modify, and consult again, involving all users at every stage of the process and listening to children, women and adolescent girls, men and boys, the elderly, people living with disabilities, and marginalized groups. As a minimum, organization staff must consult the users on design and location, management of the operation, and maintenance of activities, and they must routinely check to make sure that the services provided are being used. Community engagement efforts require skills to listen, digest, and action design considerations such as safety, privacy, accessibility, menstrual hygiene management, handwashing, operation, and maintenance.

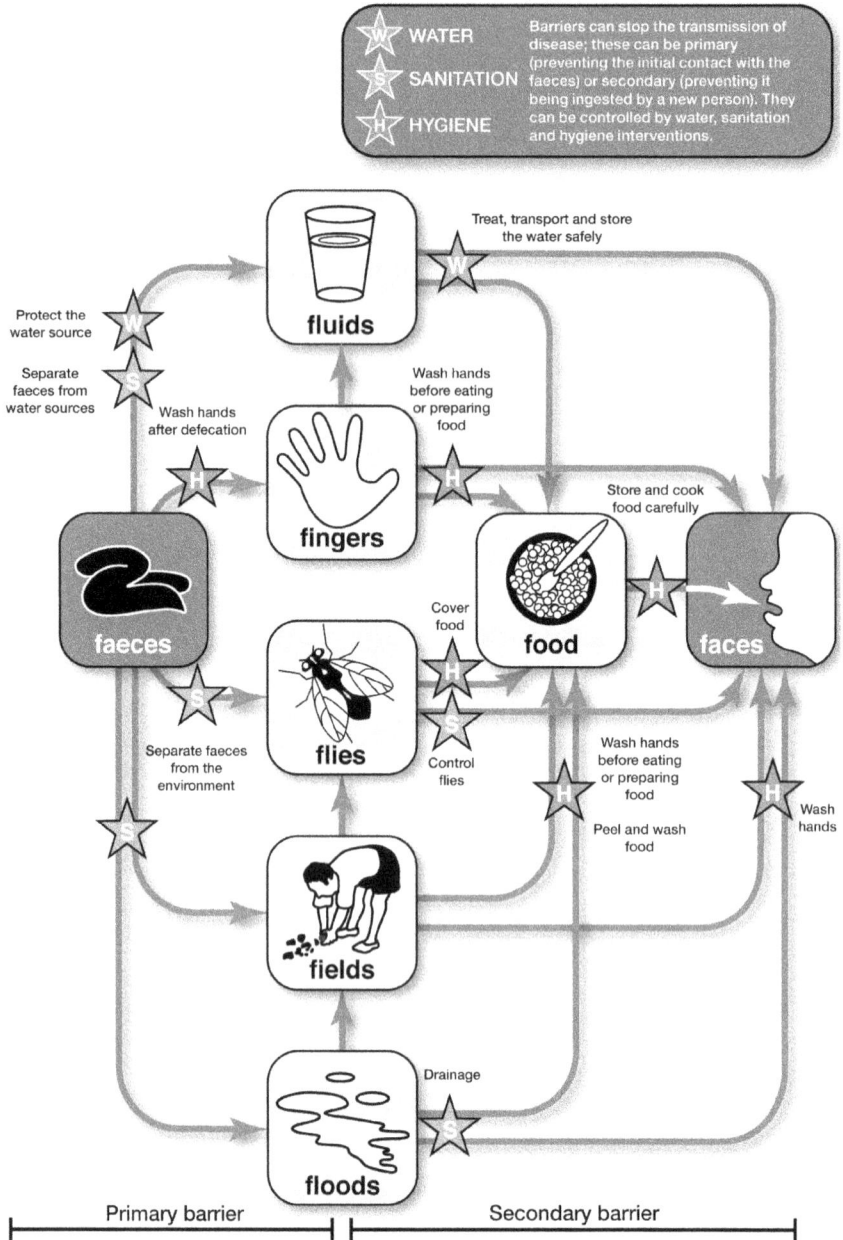

Barriers can stop the transmission of disease; these can be primary (preventing the initial contact with the faeces) or secondary (preventing it being ingested by a new person). They can be controlled by water, sanitation and hygiene interventions.

- W WATER
- S SANITATION
- H HYGIENE

Treat, transport and store the water safely

fluids

Protect the water source

Separate faeces from water sources

Wash hands after defecation

Wash hands before eating or preparing food

fingers

Store and cook food carefully

faeces

Cover food

food

faces

Separate faeces from the environment

flies

Control flies

Wash hands before eating or preparing food

Peel and wash food

Wash hands

fields

Drainage

floods

Primary barrier | Secondary barrier

Figure 2.1 Important disease transmission routes related to poor faecal sludge management

2.3 A rapid response: balancing quality and quantity

Service providers must ensure that the environment inhabited by emergency-affected populations is free from human faeces as rapidly as possible. To achieve this, they must ensure that there is immediate provision of toilet facilities within the first few days of any emergency and that any excreta found on the ground is cleared away. The immediate provision of basic excreta containment interventions is better than the delayed provision of improved systems. An environment heavily contaminated by excreta at a time when the population is at its most vulnerable can lead to the rapid onset of disease. At senior emergency management levels there will be discussions about the speed and coverage of human waste services against the quality of what is provided. This will always be a balancing act to ensure that all users have a dignified, safe, and appropriate infrastructure, both during the day and at night, as quickly as possible after the emergency has occurred.

2.4 Scaling up and monitoring

A failure to quickly scale up infrastructure construction programmes (especially of toilets) is a common problem during emergency responses. During acute emergencies involving displaced people, service providers should track not only the coverage of their services but also the rate of change of these indicators (for example, the increase in the number of toilets constructed per day) to clearly show that the target values will be met on time. If the rates show that the targets will not be met on time, additional financial, material, and human resources must be allocated immediately. Changes in implementation strategy may also be required: for example, greater use of mechanical equipment, design changes to make better use of local skills and materials, or the establishment of local factories to mass produce common components. Programmes that do not track implementation rates often fail to realize that they are failing before it is too late.

2.5 SPHERE and other faecal sludge management standards

Virtually every country in the world has laws, regulations, or codes of practice governing excreta disposal. Even though emergency relief is a special situation, it is normally illegal to fail to comply with these rules without specific written authority. Unfortunately, not only do national regulations rarely make specific reference to emergencies but they can also be a barrier to short-term solutions. This is because their rigid standards are based on normal living situations and numerical indicators and take no account of local circumstances. Non-compliance can have serious implications: local engineering colleagues could be open to prosecution and responsible institutions may refuse to approve short-term infrastructure or to take responsibility for the infrastructure when external providers leave.

In a bid to overcome these difficulties and the lack of local regulation, the international community came together to produce international norms of best practice in emergencies: *The SPHERE Handbook: Humanitarian Charter and Minimum Standards in Humanitarian Response'* (** Sphere Association, 2018) has been accepted as providing appropriate minimum standards and indicators for emergency assistance globally and forms the basis for the level of service provided by most emergency relief agencies. The SPHERE manual is an excellent reference book for anyone working in emergencies and I strongly recommend readers to have their own copy. Electronic copies are available free of charge in numerous languages from the reference given above. Parts of SPHERE will be reproduced in each chapter of this book to illustrate the level of provision expected by international standards.

When starting work in a new country or region, you should always become conversant with all relevant national and regional regulations. Unless you are told otherwise, they take precedence over any other guidance.

Where possible, compare local standards and regulations with SPHERE standards and indicators and decide what intervention is most appropriate. If SPHERE doesn't comply with local regulations, obtain formal approval before using SPHERE.

2.6   Assessment

The key to any intervention is a proper assessment of the situation and a plan of action that is developed together with users and relevant stakeholders, where possible. This is just as true for short-term immediate emergency responses as it is for long-term interventions. Only the scale and depth of the assessment changes.

Assessments are the first step in the humanitarian programme cycle and allow indicators such as those listed in the Global WASH Cluster indicator bank (Global WASH Cluster, 2009) to be addressed and ultimately measured in order to optimize impact in a coordinated manner. The assessment needs to answer the following five questions for both rapid onset emergencies and protracted crises:

1. What are the priorities to address the immediate and long-term risks?
2. Which groups are in the most need?
3. Where should interventions focus first?
4. Over what timeframe?
5. Against which standards and indicators?

2.7   Protection of water sources and the environment

It is of particular importance that immediate steps are taken to prevent the contamination of existing water sources from human excreta. During the initial stages of an emergency response, this may require the use of guards to prevent the population from defecating, bathing, or leaving their garbage close to water sources or along riverbanks. Open defecation should also be discouraged along public highways, in the vicinity of hospitals, feeding centres, reception centres, food storage areas, and food preparation areas, and in fields containing crops for human consumption. When it is impossible to establish toilet facilities quickly, managed open defecation may be necessary, but only in strictly confined areas and for a limited period. The affected population must be consulted about why such areas have been proposed and their acceptance must be obtained (see Chapter 6). Open defecation areas should be closed and covered over as soon as possible.

2.7.1 Rapid assessment

A rapid assessment should be conducted prior to any implementation and regularly thereafter to keep the organization and stakeholders up to date on what is happening. Where possible, use common data collection tools, methodologies, and indicators under the guidance of the senior agency coordinating WASH activities. This harmonized approach to assessment allows the data collected by individual organizations to be compiled in a single database and be used in shared analysis and prioritization. In major emergencies, this usually comes under the management of the WASH Cluster. Uncoordinated assessments occur where individual organizations use different indicators and methodologies; the resulting data sets cannot be combined and therefore the results cannot be used to inform the overall analysis. (For more information on this subject, see (** Davis & Lambert, 2002)).

2.8   Should we be involved?

At the completion of the assessment, the agency must ask itself, 'Should we get involved?'

- Does the agency have the right skills, appropriate personnel, and backup support?
- Does it have sufficient resources to respond at the necessary scale for the required duration?
- Does it have the appropriate beliefs, ethics, and vision to work in the scenario?

If the answer to any of these questions is 'no', then the agency should share its knowledge with organizations that do have the skills and resources to help and then withdraw. A poor intervention is usually worse than no intervention at all.

2.9   Action strategy

It takes time to develop an ideal system of faecal sludge management and so a staged/incremental approach to improvement must be adopted.

The following stages are typical and should be repeated regularly:

- Carry out a rapid survey to establish current excreta management practices, key problems, constraints, and the physical parameters of the site that might affect HWM options. Ensure that the needs/demands of vulnerable and marginalized groups (women, children, disabled, minorities, etc.) are included.
- Establish clear lines of communication with the user community, emergency response management, regulatory authorities, and partner implementing organizations. Consult regularly to ensure continuity and approval.
- Prepare an HWM plan, bearing in mind any limitations in resources, community preferences, and institutional constraints. Mobilize the necessary resources (people, money, materials, and equipment). Use a staged approach.
- Clarify roles and responsibilities, such as areas covered by intervention, operation, and maintenance, duration of responsibility, decommissioning, and financial arrangements.
- Acute response measures are likely to focus on controlling indiscriminate defecation at the earliest opportunity, particularly in critical areas, while initiating temporary measures that can be improved later.
- Stabilization measures may include the construction of communal and/or family toilets, establishing a basic solid waste management programme, or ensuring that sullage is channelled away from family homes and key institutions. Establishing a cleaning and maintenance regime is also important.

2.10   Deciding what to do

2.10.1 Rehabilitation of existing infrastructure

If existing infrastructure exists, then its rehabilitation is the first option to consider. Rehabilitation has numerous advantages, if it was an appropriate choice in the first place! Much of the original infrastructure and its links to other parts of the HWM scheme may be intact, management and operational organizations are already present and can be supported to continue providing a service, and operation and management skills and spare parts may already exist locally.

Planning the rehabilitation and reconstruction of HWM infrastructure is a task that is normally managed by specific government agencies. However, in post-disaster/emergency situations, depending on the scale of the resulting damage, aid agencies, civil society, and other organizations – both private and public – may collaborate with the government.

Rehabilitation can be a complex process that can take between a few weeks and several years. When undertaking rehabilitation programmes, it is important that the different organizations involved coordinate with the government and among themselves, and that they conform to existing national policies and standards. Links to existing long-term governmental programmes should also be examined and developed.

Once the acute needs of the affected population have been met, further assessments will indicate key HWM facilities that require rehabilitation. (adapted from **Gensch, et al., 2018).

2.10.2 *Choosing something new*

If rehabilitation is not possible or a temporary solution is needed while rehabilitation takes place, then a new solution will be required. As is clear from earlier parts of this book a large and varied range of factors must be considered, and in a very short timeframe too. In practice, this is far from simple and the task is usually given to the most experienced team members or to those who have been specially trained in the task. Numerous selection aids have been developed over the years to assist with decision making, but they have tended to oversimplify the problem and to be of only partial use in very limited situations. To date, there is no tool or application capable of taking a holistic approach to infrastructure selection.

2.11   Ownership

Unless agreed otherwise in writing, organizations and agencies that install any infrastructure own it!

This is very important because it establishes who is responsible for the infrastructure's operation, maintenance, and possibly eventual removal. This is particularly true of infrastructure constructed for use by individual families or small communities. Just because you have provided, say, a toilet for a family doesn't mean they own it; you still do so unless there is a clear formal agreement that states otherwise. This also applies to public infrastructure such as garbage collection or sullage drainage. There is no obligation on public bodies to take responsibility for infrastructure unless this has been agreed formally, preferably at the very beginning.

CHAPTER 3

Non-technical issues affecting service delivery and sustainability

For the purposes of this book, a sustainable intervention is one that meets the needs of all of the people it is intended to assist for the whole time over which it is expected to function. This includes all elements related to the planning, design, construction, operation, and maintenance of the intervention. Service delivery is part of sustainability but with a greater focus on the successful delivery and operation of the intervention.

There are many issues that must be taken into account when delivering a good, sustainable service; the main ones are discussed in this chapter and Chapter 4 and they must be borne in mind when using the rest of the book.

3.1 Faecal sludge management standards and assessment

SPHERE begins by setting out the justification for excreta management:

An environment free of human excreta is essential for people's dignity, safety, health and well-being. This includes the natural environment as well as the living, learning and working environments. Safe excreta management is a WASH priority. In crisis situations, it is as important as providing a safe water supply.

All people should have access to appropriate, safe, clean and reliable toilets. Defaecation with dignity is a highly personal matter. Appropriateness is determined by cultural practices, people's daily customs and habits, perceptions, and whether individuals have used sanitation facilities before. Uncontrolled human defaecation constitutes a high risk to health, particularly where population density is high, where people are displaced, and in wet or humid environments.
(** Sphere Association, 2018).

Excreta management standards and key indicators are given in Box 3.1. Despite the comprehensive detail provided in SPHERE, care must be taken when applying numerical recommendations. No context or population is homogeneous: people have different genders, ages, sizes, levels of mobility, customs, and beliefs, and there is a diverse range of opportunities and challenges with stakeholders at the national, regional, and global level. Indicators are not standards, they are suggestions of what is needed to meet the standards. That means that they should change to suit the requirements of individual situations. SPHERE focuses its advice and guidance on providing users with hygienic and dignified facilities in which to excrete (i.e. toilets). SPHERE is weak on faecal sludge management (FSM) chains (i.e. beyond the toilet and on-site storage) and so local standards and regulations will normally be more informative.

SPHERE also includes a good example of a checklist for assessing excreta management needs (Box 3.2), although, again, the list is limited in terms of advice on off-site excreta management.

Box 3.1 SPHERE excreta management standards and key indicators

Excreta management standard 3.1: Environment free from human excreta

All excreta are safely contained on-site to avoid contamination of the natural, living, learning, working and communal environments.

Key indicators

- There are no human faeces present in the environment in which people live, learn and work.
- All excreta containment facilities are sited appropriately and are an adequate distance from any surface or groundwater source.

Excreta management standard 3.2: Access to and use of toilets

People have adequate, appropriate and acceptable toilets to allow rapid, safe and secure access at all times.

Key indicators

- Minimum 1 toilet per 50 people rapidly improved to 1 toilet per 20 people.
- Maximum walking distance of 50 metres between a dwelling and a shared toilet.
- Percentage of toilets that have internal locks and adequate lighting.
- Percentage of toilets reported as safe by women and girls.
- Percentage of women and girls satisfied with the menstrual hygiene management options at toilets they regularly use.

Excreta management standard 3.3: Management and maintenance of excreta collection, transport, disposal and treatment

Excreta management facilities, infrastructure and systems are safely managed and maintained to ensure service provision and minimum impact on the surrounding environment.

Key indicator

- All human excreta are disposed of in a manner safe to public health and the environment.

Source: (** Sphere Association, 2018).

Box 3.2 Emergency excreta management initial needs assessment checklist

General data

- How many people are affected and where are they? Disaggregate the data by sex, age, disability and so on.
- What are people's likely movements? What are the security factors for the affected people and for potential relief responses?
- What are the current, prevalent or possible WASH-related diseases?
- Who are the key people to consult or contact?
- Who are the vulnerable people in the population and why?
- Is there equal access for all existing facilities, including at public places, health centres and schools?

(Continued)

Box 3.2 Continued

- What special security risks exist for women, girls, boys and men? At-risk groups?
- What water, faecal management and hygiene practices were the population accustomed to before the crisis?
- What are the formal and informal power structures (for example, community leaders, elders, women's groups)?
- How are decisions made in households and in the community?
- Is there access to local markets? What key WASH goods and services were accessible in the market before the crisis and are accessible during the crisis?
- Do people have access to cash and/or credit?
- Are there seasonal variations to be aware of that may restrict access or increase demands on labour during harvesting time, for example?
- Who are the key authorities to liaise and collaborate with?
- Who are the local partners in the geographical area, such as civil society groups that have similar capacity in WASH and community engagement?

Excreta management

- Is the environment free of faeces?
- If there is open defecation, is there a designated area?
- Are there any existing facilities? If so, are they used? Are they sufficient? Are they operating successfully? Can they be extended or adapted?
- Are the facilities safe and dignified: lighted, equipped with locks, privacy screens? Can people access the toilet facilities during the day and night? If not at night, what are the alternatives?
- What excreta management practices does the host population practice?
- Is the current defecation practice a threat to water supplies (surface or groundwater) or living areas and to the environment in general?
- Are there any social–cultural norms to consider in the design of the toilet?
- Are people familiar with the design, construction and use of toilets?
- What local materials are available for constructing toilets?
- Is there an existing acceptance of and practice for composting?
- From what age do children start to use the toilet?
- What happens to the faeces of infants and young children?
- What is the slope of the terrain?
- What is the level of the groundwater table?
- Are soil conditions suitable for on-site excreta disposal?
- Do current excreta disposal arrangements encourage vectors?
- Are there materials or water available for anal cleansing? How do people normally dispose of these materials?
- Do people wash their hands after defecation and before food preparation and eating? Are soaps or other cleansing materials with water available next to the toilet or within the household?
- How do women and girls manage menstruation? Are there appropriate materials or facilities available for this?
- Are there any specific facilities or equipment available for making toilets accessible for persons with disabilities, people living with HIV, people living with incontinence or people immobile in medical facilities?
- Have environmental considerations been assessed: for example, the extraction of raw materials such as sand and gravel for construction purposes, and the protection of the environment from faecal matter?
- Are there skilled workers in the community, such as masons or carpenters, and unskilled labourers?

(Continued)

> **Box 3.2** Continued
> - Are there available pit emptiers or desludging trucks? Currently, is the collected faecal waste disposed of appropriately and safely?
> - What is the appropriate strategy for management of excreta – inclusive of containment, emptying, treatment and disposal?
>
> *Source*: (** Sphere Association, 2018)

3.2   Public health risks to faecal sludge workers

Faecal sludge managers must ensure that an assessment of public health risks is carried out for all activities related to FSM and a plan is in place to mitigate and monitor priority risks. All those involved in FSM-related activities must have the relevant PPE and training for the tasks they are performing. Personnel undertaking the following activities should be prioritized:

- Anyone involved in the direct handling of human excreta, solid waste or sullage, including desludging, collection, movement, treatment, and disposal;
- Anyone involved in the maintenance or cleaning of human related wastes infrastructure, such as vehicles, machinery, equipment, tools, or clothing.

The best way to undertake a public health and safety risk assessment is to analyse risks at each step of the management chain, from source to final use/disposal. At each step, risks should be analysed in terms of probability of contact with human related wastes (rated from high to low) and likely impact to health (rated from high to low). More information can be found in the UNHCR's *Operational Guidelines for Staff* at (UNHCR, 2015b)

3.3   Social and cultural considerations

3.3.1 Community involvement

All evidence shows that the most common reasons for the failure of FSM facilities are not technical but social and institutional. Failure to address these issues from the very beginning of an intervention will seriously affect the likelihood of a successful outcome (Box 3.3).

   The delivery and management of toilets and associated facilities require the cooperation of users to ensure that they are used hygienically and are kept clean and functional. Therefore, as soon as possible, it is important to establish some form of communication with the community, for instance by establishing a sanitation committee. The committee can be formed using the community contacts established during the surveys and should include community representatives, agency staff, government personnel, and groups involved in hygiene promotion. In a temporary settlement, the sanitation committee will be one of several camp committees. In a large community, or in a community comprised of different groups of people, it may be necessary

> **Box 3.3** Women's use of emergency toilets
>
> Recent research has shown that, on average, 40 per cent of women are not using the toilets provided. The main reasons for this are:
>
> Not wanted to be seen going to the toilet.
> Sexual harassment.
> Lack of locks on doors.
> Lack of privacy.
> Lack of lighting at night.
> Fear of vermin.
> Cleanliness.
>
> Building toilets that are not used is a waste of time, money, and resources. More importantly, we are failing the community we are trying to help and increasing the public health risks to individuals and to the community at large.
>
> Consultation with the user community at all stages in toilet provision cannot be overstated (Oxfam GB, 2018).
>
> **CONSULT → MODIFY → CONSULT**

to establish a number of sanitation sub-committees. Sanitation committees commonly have the following responsibilities:

- To raise awareness of the importance of effective faecal sludge services.
- To feedback the issues and views of the community to the delivery team related to faecal sludge services.
- To promote appropriate hygienic practices (handwashing, proper disposal of infants' and toddlers' excreta, etc.).
- To promote the proper use of facilities and their positive benefits to families and the community.
- To consult on the best faecal sludge options.
- To recruit and organize people for toilet construction, supervision, maintenance, and management.
- To provide regular monitoring of the faecal sludge measures.

In most cases, communities will be most concerned about the facilities that directly affect them, particularly the location, design, and quantity of toilets; community consultation and cooperation in these matters is therefore essential. Details of the rest of the FSM chain and medium- to longer-term solutions are also relevant to the community and must be carefully planned with users.

If it is clear that the timeframe of the humanitarian situation will be longer than six months, the best guarantee that people will safely defecate in a toilet and that toilets are kept clean and functional is to encourage the construction of shared or individual household toilets as quickly as possible. Toilet providers should ensure that every household that wishes to construct their own toilet can access tools, materials, and technical support to build, operate, and maintain an appropriate facility from the outset of the emergency. Toilet providers should provide one household toilet-digging kit to be shared

---

Box 3.4 Family toilet-digging toolkit

General information and description

High-quality and robust hand tools for accelerating family toilet construction activities. The kit can also be useful for defecation fields, clean-up campaigns, or preparation of all types of groundworks, including toilet cubicles, pits, trenches, foundations, and road repair.

Kit contents

Quantity	Item
10	Spades with square mouth and straight wooden shaft with handle.
10	Shovel with round mouth and straight wooden shaft with handle.
10	Pickaxe 3 kg head, chisel, and point with hardwood shaft.
6	Bucket, 16–20 litre, galvanized steel or heavy-duty plastic.
6	Rake, 12 tooth head, blade width 208 mm, with wooden handles and screw fixing.
5	Steel bar with chisel and point ends, 30 mm diameter, 1.5 m long.
5	Builder's chisel, 225 mm long and 75 mm wide blade.
5	Lump hammer, 2 kg.
1	Measuring tape, 30 m long.
2	8 mm polypropylene rope, 30 m long.
1	Site marking tape, red/white, 500 m long.

Weight, volume, and packaging

Gross weight	152 kg.
Packaging	Sturdy timber frame crate suitable for sea/air freight.
Dimensions	1,480 × 600 × 630 mm.

Source: Adapted from. (UNHCR, n.d).

---

between every 10 families and provide them with technical support where necessary. The suggested contents of the kit are shown in Box 3.4. In the cases where families cannot provide their own materials for a toilet, toilet providers may enable access (either in-kind, cash or voucher system via the local market) to items such as toilet slabs and cubicle material to facilitate construction of household toilets. Dedicated staff for the household toilet programme should be recruited and mobilized as soon as is possible.

No one can force an individual to use a toilet; they have to want to. Similarly, attitudes and practices around handwashing also vary widely. We must recognize that we are all different and there is no single toilet design that will suit everyone. Gender, age, mobility, traditional practices, beliefs, and aspirations all affect the acceptability and impact of excreta disposal methods. Close collaboration with user communities throughout all stages of service delivery is critical to ensuring that the facilities provided are appropriate and acceptable. However, in major emergencies where populations are traumatized, it may be possible, through good hygiene promotion, to change perceptions of what is acceptable hygiene practice even during the onset of an emergency. As an example, selecting appropriate places for defecation is a compromise between the wishes of the users, regulations, and technical feasibility. In many cases, options may be very limited, especially in the early stages of the response, but close consultation with users is essential if the toilets that are constructed are to be used. More information on this is given in Box 5.1.

3.3.2 Personal security

Personal security is a primary consideration when wishing to visit a toilet, especially one used by a general population. This is particularly true for women and young children and especially at night. There are various ways in which these fears can be minimized:

1. Provide family or small group toilets. This shortens the distance users must travel to get to/from the toilet and limits the number of other people with whom they may come in contact.
2. Provide good-quality access paths (Table 5.1) and artificial lighting along main walking routes, in the public areas of a toilet block, and in the toilet cubicle. Modern LED lighting has very low power consumption, making good lighting more affordable. Alternative personal lighting such as flashlights could be provided. Where lighting is difficult, discuss the issue with community members and arrange for vulnerable people to be chaperoned. In any case, lighting should **always** be provided in the cubicles that are expected to be used at night. Further information on lighting can be found on the Oxfam WASH website (Oxfam GB, 2018).
3. Communal toilet blocks should always be supervised at night, preferably by persons who are trusted by the user community. Where possible, female attendants should be used.
4. The walls and doors of toilet cubicles should be constructed of a solid material such as wood or cement blocks to minimize the risk of someone breaking into the cubicle while it is in use. This is difficult to achieve during the acute phase of a response because of the need for speed of construction and the flexibility of demand. Additional attendants should be employed to ensure that users feel safe while visiting the toilet.
5. Toilet doors should be strong and large enough to give the users privacy. They should also be fitted with a lock that can be used from the inside. See Chapter 5, section 2.1 for more details.
6. Toilets must be clean and free from vermin and, as much as possible, flies. There is more on this subject in Chapter 6, section 4.

Despite all these interventions to improve safety and security, there may still be some who are unwilling to use the toilets provided. For them, there are two possible approaches: work closely with individual families to try to understand why they are unwilling to use the facilities and develop tailored solutions to overcome the issues; or provide individual users with the equipment to establish a toilet in the home (such as a commode).

3.3.3 Child excreta management

Hygiene promotion measures for children's faeces include understanding the needs and preferences of mothers/caregivers and their children and the availability of items via local markets. Once these issues are understood, information and training to mothers/caregivers about safe disposal options, children's toilet training, and laundering practices can be provided and active advocacy can be undertaken to prevent indiscriminate defecation and household contamination with children's faeces. Hygiene promotion includes hygiene messages on the importance of handwashing with soap after contact with child excreta and washing the child after defecation. It may also include encouraging the clean-up of already contaminated environments with shovels or other tools to avoid direct contact with children's excreta.

Children's faeces are generally more dangerous than those of adults because excreta-related infections are more prevalent in children, with a higher prevalence of diarrhoea and soil-transmitted helminth infections. In addition, toddlers and small children are often unable to fully control their defecation and may defecate in areas where other children could be exposed. Hence, children are more susceptible to faecal–oral transmitted diseases. In some societies, children's faeces are considered harmless and are not collected or disposed of safely. Additionally, children often do not use a toilet because of their age, stage of physical development, or the safety concerns of their parents. Addressing child excreta management therefore includes the context-specific consideration of the following components:

- Children may not be happy using toilets that have a pit below them for fear of falling down the hole, and so an alternative defecation place should be made. One solution is to include a small toilet interface in the design. Children generally do not require the same privacy as adults, and they are likely to prefer to defecate outside where they can be seen if they have a problem. Parents may also be reassured as children will always be in view.
- Public and shared toilets should be close to households, have proper lighting, and be equipped with child-friendly user interfaces such as smaller bowls or squat holes. The cubicle has to be large enough to be occupied by a caregiver and child together. A children's toilet can be further enhanced with child-friendly, colourful artwork and picture-based hygiene messages.
- Age-appropriate faecal containment products such as nappies, diapers, and potties need to be considered. If disposable nappies or diapers are being used, there needs to be an adequate collection and management system in place with subsequent burial or treatment options. Washable nappies may be an alternative, assuming that there is a place to wash and dry them and soap or other products to ensure that they are sanitized. The wash water should also be disposed of in a toilet. If potties are being used, the child faeces can be discarded or rinsed into the toilet and the potty cleaned with soap or household bleach afterwards. Disposable nappies, nappy liners, and 'wet wipes' should never be placed in a toilet but tightly sealed in a plastic bag and disposed of with the domestic garbage. (Section adapted from ** Gensch, et al., 2018).

3.3.4 Anal cleansing

Water, paper, stones, maize cobs, and other materials are used for cleaning the anus after defecation. While it is important to recognize what people use traditionally, it may be necessary to encourage people to use more appropriate materials such as paper or water. Non-biodegradable materials for anal cleaning should be strongly dissuaded as their use can cause supply and disposal problems. They also have a marked effect on the life of the toilet and the operation of the rest of the FSM chain. Where the use of non-biodegradable materials cannot be prevented, facilities for their separate collection and disposal must be provided next to the toilet interface. Generally, this will be a watertight container, preferably with a foot-operated lid. The contaminated materials should then be buried or burned daily and not deposited where they will create a health or environmental hazard. Staff tasked with emptying anal cleansing materials should be given appropriate PPE and training for the task.

Users should be encouraged to provide their own anal cleaning materials; however, where that is not possible, it is the responsibility of those who construct the toilets to ensure their ready availability. Where water is used for anal cleansing, toilet providers must ensure that an adequate water supply and suitable washing containers are always available. At least 2–3 litres of anal cleansing water should be planned for per person per day. Reservoirs for anal cleansing water should be covered to prevent vector breeding.

Budgeting for demand from communities who prefer to use toilet paper is very difficult as usage varies widely, as does the size of a toilet roll. Guidance on calculating the quantity of toilet paper required is shown in Box 3.5. Toilet paper is a heavily traded product so very careful management is required to prevent theft by workers and users.

Box 3.5 Calculating toilet paper requirements

Calculating toilet paper requirements is very difficult. The data provided below may help with preliminary planning but will almost certainly have to be adapted as experience is gained.

A commonly quoted figure is that, on average, each person uses 57 sheets (or 60 sheets) of toilet paper each day.

As a starting point, this figure could be used to estimate the total number of sheets required thus:

Total number of sheets (S) = 57 × Number of users (P) × Number of days' supply required (T)

Converting this to a bulk number of toilet rolls or weight is very difficult because of the variety of roll sizes, paper weight, number and size of sheet, and numbers of layers (ply) of paper in a sheet. One method could be to get suppliers to quote based on the total number of sheets required, but a close examination of the quality and presentation of the paper would be required prior to purchase.

Another method is to estimate the weight of toilet paper required. On average, an individual uses 5.2 kg of toilet paper each year (Morris, 2021) (approximately 15 g per day), although that, too, varies widely.

(Continued)

> **Box 3.5** Continued
>
> **Calculation**
>
> 1. Assume a daily per person weight of toilet paper used (A) e.g. 15 g
> 2. Estimate the number of users each day (P) e.g. 200
> 3. Decide the number of days' supply required (T) e.g. 30
> 4. Total weight (kg) of toilet paper required = A × P × T/1,000 15 × 200 × 30/1,000 = 90 kg

3.3.5 Menstruation and incontinence

From day one, enabling women and girls to practise appropriate menstrual hygiene management helps them to live with dignity and engage in daily activities (school, market, etc.). The use of inappropriate materials or the provision of insufficient facilities for the use and disposal of menstrual hygiene products can quickly become a serious problem. It is an important issue that must be addressed sensitively.

Similarly, many people either occasionally or frequently suffer from incontinence. Incontinence may not be a widely used term in some contexts, even within the medical profession. It is a complex health and social issue that occurs when a person is unable to control the flow of their urine or faeces. It can lead to a high level of stigma, social isolation, and stress and an inability to access services, education, and work opportunities. Prevalence may seem low, as many people will keep it a secret, yet a wide range of people may live with it. Poor incontinence hygiene management can be a major source of disease transmission in emergencies (** Sphere Association, 2018).

Comprehensive guidance on the integration of menstrual hygiene management (MHM) can be found in (Sommer, 2017). The key actions for MHM and incontinence in emergencies include ensuring that toilet facilities enable private, dignified, and user-friendly changing, laundering, and disposal of menstrual and incontinence protection materials. It is critical to ensure that women, adolescent girls, and those suffering from incontinence are involved in the design. Usability features can include easy access to water, soap, and replacement menstruation or incontinence material. People with incontinence and their carers each need five times as much soap and water as others. In addition, the toilet cubicle should be provided with hooks to hang up possessions and small shelves on which to put pads, soap, and other items. Options may be required for laundering sanitary protection or incontinence materials. This may require installing collection buckets in each toilet and bathing unit cubicle. Persons suffering from acute incontinence may not be able to visit a toilet; in those cases, suitable sanitary facilities should be provided for use in the home (see Chapter 5, section 2 for more details).

Used sanitary protection and incontinence products contain blood and/or faeces and their disposal could pose a high risk for the transfer of infectious diseases. In general, they should not be disposed of in the toilet as they may cause problems with all elements of the FSM chain. A plastic bucket with a lid should be provided in every cubicle used by females and their use encouraged. The buckets should be emptied daily and sterilized before being returned to service. Alternatively, the buckets can be provided with a disposable liner that is replaced daily. However, this approach may lead to problems in the solid waste management chain (see Chapter 15). Disposal commonly takes the form of burial in a separate pit or via other recognized disposal facilities that do not allow contamination of the operation and maintenance of drinking water sources.

3.4   Institutional and political environment

All emergency relief takes place within the wider national and international environment and the impact of that environment on decision making cannot be ignored. The most important issues are discussed below, but for more detailed information see (** Davis & Lambert, 2002).

3.4.1 Institutions

Very few countries have a ministry or department specifically responsible for all aspects of FSM; it is usually divided between a number of agencies at national, regional, and local level. A general understanding of each agency's roles and responsibilities will identify which ones are the most important to both the emergency and long-term response. Dialogue with these agencies is necessary because they will be responsible for policing the regulations and be most likely to take over long-term responsibility for operation and maintenance. An understanding of their technical, managerial, and financial competences will have an impact on the range of technologies considered.

3.4.2 Politics

The role of politics and politicians in emergency relief cannot be ignored. Their views and influence will have a direct bearing on the technology choice, especially in the early stages of the response. It is very common for political leaders to assume that the impact of an emergency will be short-lived and only temporary measures should be adopted as the crisis will soon pass and all the emergency infrastructure will be demolished. Unfortunately, evidence suggests that this is rarely the case, and this contradiction can lead to difficulties in service delivery.

There is also the contentious issue of comparability of service level. Delivering a level of service (such as the number and quality of toilets) to displaced people that is better than that of the host population quickly leads to friction between the two communities. Local authorities may be unwilling to support or approve the services proposed or to take responsibility for them at the end of the intervention. An understanding of the wider social and political environment is very important, especially when undertaking a large programme. In major emergencies, it is usual for a neutral agency such as UNHCR or UNICEF to act as a go-between; they should ensure that activities offered to displaced people are acceptable by the wider community. However, in smaller emergencies it may be up to the implementing organization to take on that role.

In urban areas where service levels are often higher, the government is more present, networks are more established, and service provision is often through utilities, municipalities, or private companies, technology choice will often be driven by these stakeholders. However, simple acute phase responses (e.g. trench toilets or container toilets), sometimes thought of as 'rural technology', are often the best short-term solution. Careful negotiation with the key stakeholders is normally necessary to gain their acceptance. In rural areas where populations remain disbursed, centralized systems such as piped sewerage are seldom useful and household toilet options are generally more appropriate. Local and national stakeholders are normally more accepting of these choices in these circumstances.

3.4.3 Financial sustainability

The respective costs for the long-term operation and maintenance of human waste infrastructure need to be considered during technology selection. While cost recovery is not a priority in an acute humanitarian response, awareness of the protracted financial consequences of (re-)establishing human waste management (HWM) services is essential from the outset.

To function in the long term, post-emergency solutions must be financially affordable for the local population. If expensive or over-complex solutions are deployed in the immediate emergency phase, emergency FSM organizations must ensure that clear strategies are developed from the start for transferring to more efficient and sustainable technologies, particularly those with reduced dependence on imported equipment, fuel supplies, or chemicals. The cost per beneficiary of excreta management services should be tracked to ensure that they can be adapted to a cost level that is functional yet affordable for the local community.

3.5 Environmental impact

Environmental impact is becoming increasingly important, not only due to the high demand for environmental resources for toilet construction (particularly wood) but also because of environmental pollution and climate change.

Poorly managed excreta management activities in emergency settings can have an extremely negative impact on the environment. Pollution of surface water resources with nitrate-laden faecal sludge may lead to irreversible impacts on fragile ecosystems. It may also have an impact on groundwater quality, but this is often overstated (see Chapter 4, section 4).

Unsustainable and inappropriate procurement of river sand/gravel or wood or burning bricks for toilet construction may present a significant environmental risk, particularly in locations where sustainable supplies of raw materials such as wood are in limited supply. The risk could include landslides, flooding, and consequences for the flora, fauna, and aquatic environment. When sourcing construction materials, ensure that you visit the site where providers source their materials to check that the location and means of collection are responsible and appropriate.

Emergency programmes should ensure that all excreta, blackwater (from septic tanks), sewage, faecal sludge, solid waste and sullage are treated, conveyed, reused, or disposed of in a manner that minimizes any potentially negative impacts on the environment. In all settings, they must not be allowed to enter the environment without treatment.

National standards, sanitary codes, and environmental measures must be respected once emergency standards have been met. Faecal wastes of environmental and public health importance must be dealt with immediately; wastes that are comparatively inert and require a longer-term approach are less time critical.

During short-term emergency responses, it may be possible to negotiate with national authorities for environmental standards to be temporarily relaxed. In cases where environmental legislation does not exist, then the minimum guidance provided in SPHERE must be aimed for (** Sphere Association, 2018). Programmes should be designed and monitored in close collaboration with local regulatory authorities. All environment-related rules, regulations, and norms should be communicated clearly to the population.

In all cases, acute-, stabilization-, and recovery, environmental impacts from interventions must be considered from the outset of the emergency. Failure to do so can have widespread ramifications and prove costly to address. Preventative and mitigation measures are far more cost-effective than remedial actions. Environmental measures must be budgeted from the start.

3.6   Security and safety

Toilet providers must take every precaution to protect and safeguard populations and ensure that toilet facilities minimize the threats to users, especially women and girls, day and night. Lack of security, or dignity, when using facilities can lead to anxiety concerning their use. Inappropriately designed or located toilet facilities can lead to a risk of gender-based violence: for example, poor facilities may force women and girls to defecate indiscriminately, putting them at greater risk. Focus group discussions and key informant interviews should not only focus on what needs to be done to make facilities safer but should also aim to understand why sections of the population may be at risk so that toilet facilities can be adapted to reduce that risk. Always feedback the outputs from community consultation so that people can see the impact of their participation. Advice on how to put these ideas into practice is given in Chapter 5.

Due to the security risks associated with public toilets, FSM programmes should start with or transition as quickly as possible into the construction of household toilets, or toilets shared between families. This approach is essential if it is clearly the cultural preference, the timeframe of the humanitarian situation will undoubtedly be longer than six months, or there are major protection issues with communal facilities. The best guarantee that people will feel safe and will defecate with dignity, and that toilets will be kept clean, is to encourage the construction of shared or household toilets as quickly as possible. However, public toilets will always be needed for public places, institutions, etc., so their location, design, and operation must be designed carefully in discussion with users to minimize the security and safety issues for all potential users.

3.7   Exit strategy

An exit strategy in the context of FSM is a planned approach to why, what, when, and how implementing organizations will end their emergency engagement. This includes the process of transitioning, handing over and decommissioning infrastructure, and exiting or disengaging from activities, projects, programme areas, or countries.

Potential exit and transition strategies should be considered from the start of activities and should be implemented as soon as basic HWM services are (re-)established to a level that successfully reduces vulnerabilities brought about by acute environmental health risks. These criteria help compare the advantages and cost-effectiveness of a sustained humanitarian intervention with those of an intervention led by local authorities and agencies, or by other donors and/or partners. Exit and transition strategies are context dependent. However, they must be addressed at an early stage of an intervention for reasons of transparency with partners and to promote a seamless handover to respective government departments or humanitarian/development partners. Humanitarian HWM interventions must be in line with national strategies and policies. If the local situation allows, interventions should be carried out in coordination with the government, local communities, and/or relevant development actors to jointly define their scope and focus. Implementing partners must specify when and how project support will be terminated and handed over to the local government, other local organizations, or service providers capable of sustaining or maintaining the achieved service levels, or they must clarify whether and how projects will be followed up (e.g. by another phase and the potential for follow-on funding to continue HWM activities if needed).

Many agencies have their own guidelines for exiting: a good example is provided by Oxfam (Oxfam WASH, n.d.). Further practical details of decommissioning HWM infrastructure are given in Chapter 10, section 7. This section is largely adapted from ** Gensch, et al., 2018.

CHAPTER 4

Core technical design parameters for faecal sludge management

The technical design of human waste management services can be undertaken only with a full understanding of the non-technical environment in which the work is to be carried out. These issues are set out in Chapter 3.

4.1   Technical sustainability

Faecal sludge management (FSM) interventions must support locally appropriate technologies and designs as well as using available and affordable local construction materials. Interventions need to be balanced between technically feasible solutions and what the affected population, local government entities, or service providers want and can manage after the project ends.

Excreta management providers should ensure that all solutions and hardware are appropriate for the local technical and financial conditions. Designs should reflect the norms and traditional practices of the local population while respecting public health best practice. Technologies should be simple enough to be operated and maintained by local service providers or by the emergency-affected population themselves with limited external assistance.

If technologically advanced solutions are deployed in the immediate emergency phase, clear strategies must be developed from the start that include transfer to more efficient and sustainable technologies, reduced dependence on imported equipment, fuel supplies, or imported chemicals, and increased ownership. These ideas are summarized in Box 4.1.

4.2   Space requirements

Excreta management requires a lot of space when fully operational – not just the land required to construct toilets, but also the requirements for sludge transfer, waste treatment, and waste reuse/disposal. Not all of this land will be required immediately and some of it may be away from the emergency area, but it needs to be identified and assigned as early as possible to prevent other services or the affected population using it. There is remarkably little information on space requirements available, but the data shown in Table 4.1 is a starting point.

> **Box 4.1** Considering sustainability in human waste management rehabilitation programmes
>
> - Avoid building infrastructure that is exposed to hazards, is inefficient, or is too small for the demand.
> - Ensure technical sustainability – local technical capacity and available materials should match the level required by the technology being implemented.
> - Build on local knowledge and utilize local materials, skills, and labour where appropriate and possible.
> - Where local communities or government are to operate and maintain the infrastructure, they should be involved throughout entire project cycle.
> - Where required, increase the community's and local authorities' knowledge and capacities on the operation and maintenance of the infrastructure that they will eventually take over.

Table 4.1 Area requirements for the different elements of emergency excreta management

Element	Area required
Simple family toilet (inc. handwashing)	2–4 m^2
Six-compartment communal toilet block	15–40 m^2
Faecal sludge storage/transfer station	60–100 m^2 per 10,000 litres
Faecal sludge treatment	0.011–0.15 m^2 per person served (OCTOPUS, n.d.)^
Disposal/reuse	No data

Note: ^Maximum and minimum space used by FSM treatment plants in Cox Bazar and Rakhine State in 2019–20. Excludes one plant at 0.68 m^2. Some sites may include land for disposal/reuse.

4.3   Volumes and characteristics of stored human excreta and faecal sludge

Faeces, urine, water, anal cleansing material, and other waste materials are commonly deposited in toilets and frequently into some form of storage container below them (such as a pit). In an emergency, toilets will be used intensively and short-term accumulation rates in the container will be high. In the longer term, an increase in the number of toilets, biological activity within the container, consolidation, and sometimes infiltration all reduce the volume of sludge. Generation and accumulation rates vary widely around the world, but, in the absence of local information, the figures given in Table 4.2 are suggested.

Making arrangements for the separate collection of non-faecal waste and establishing an effective solid waste management programme both make a major difference to the storage volume required. If the toilet design relies on infiltration of liquids into the soil, the accumulation rate will be considerably lower.

4.4   Soil infiltration and groundwater contamination

A reliable knowledge of existing soil and groundwater conditions is important, especially when infiltration-based FSM systems such as pit toilets and tertiary effluent treatment disposal are being considered. The two main concerns are

Table 4.2 Faeces and urine characteristics and suggested maximum sludge accumulation rates

Characteristics	Value suggested for emergencies	Range
Weight of wet faeces	250 g/person/day – low-income countries	51–796 g/person/day
	126 g/person/day – high-income countries	75–374 g/person/day
	85 g/person/day – infants (1–4 years)	
	100 g/person/day – children (3–18 years)	
Dry weight of faeces	38 g/person/day – low-income countries	
	28 g/person/day – high-income countries	
Defecation frequency	1.1 defecations/person/24 hours	0.74–1.97 defecations/person/24 hours
	(60% between 6 a.m. and 10 a.m.)	
Volume of urine	1.4 litres/person/day	0.6–2.5 litres/person/day
Number of urinations per day	6 urinations/day	2–11 urinations/day
Short term accumulation rate of faeces^ ^^	0.25 litres/person/day – low-income countries	
	0.13 litres/person/day – high-income countries	
Short-term accumulation of urine and anal cleaning water	2.6 litres/person/day	
Short-term accumulation of bathing water	6.7–8.7 litres/person/year	
Excreta retained in water where degradable anal cleaning materials are used	40 litres/person/year	
Wastes retained in water where non-degradable anal cleaning materials are used	60 litres/person/year	
Excreta retained in dry conditions where degradable anal cleaning materials are used	60 litres/person/year	
Wastes retained in dry conditions where non-degradable ana cleaning materials are used^^^	90 litres/person/year	

Notes: ^ For waste storage in sealed containers, use the short-term accumulation rates for design purposes.
   ^^ Short-term accumulation rates will greatly increase if solids or paper are used for anal cleansing.
   ^^^ In very dry conditions, such as arid climates, or in well-ventilated pits, any form of paper used for anal cleaning is unlikely to degrade, even after an extended time. It can quickly make up the major component of a pit's contents.
Source: Adapted from (Franceys, 1992), (Rose, 2015)

Table 4.3 Recommended infiltration rates for different soil types

Soil type	Description	Clean water (Litres/m²/day)	Filtered wastewater (Litres/m²/day)
Fissured	Variable	Highly variable^	Highly variable^
Gravel and coarse and medium sand	Moist soil will not stick together	720–2,400	50
Fine and loamy sand	Moist soil sticks together but will not form a ball	480–720	33
Sandy loam and loam	Moist soil will form a ball but still feels gritty when rubbed between the fingers	480–720	25
Loam and porous silt loam	Moist soil forms a ball that easily deforms and feels smooth when rubbed between the fingers	240–480	20
Silty clay loam and clay loam	Moist soil forms a strong ball that smears when rubbed but does not go shiny	120–240	10
Clay	Moist soil moulds like plasticine and feels very sticky when wet	24–120	<10

Note: ^Effluent infiltration systems are not generally recommended in fissured or coarse gravel soils as the effluent will travel considerable distances untreated.
For more information on assessing soil texture, see Appendix 1.
Source: Adapted from (** Davis & Lambert, 2002)

that the soil will be unable to dispose of the entire liquid part (effluent) of the wastes and that the effluent will percolate through the soil into underlying groundwater and cause bacteriological and chemical contamination. The physical, chemical, and biological processes that occur when effluent passes through soil are extremely complex and difficult to predict, but, for design purposes, the figures given in Table 4.3 suggest the maximum infiltration rates to be used in different soil types.

In dry soils (except in fissured rock), effluent normally travels vertically with very little horizontal dispersion. When pollutants enter the water table, they travel in the direction of groundwater flow (Figure 4.1) until they eventually die off or are diluted to a level that is no longer a health issue. Ground and groundwater contamination is a health risk only when it affects a drinking water source; it is not a problem in itself. There are three ways of reducing that risk:

- vertical separation between the bottom of the pollution source and the groundwater table;
- increasing the pollution travel distance within the groundwater between the pollution source and the water supply point;
- vertical separation between the pollution source and the water abstraction point below the water table.

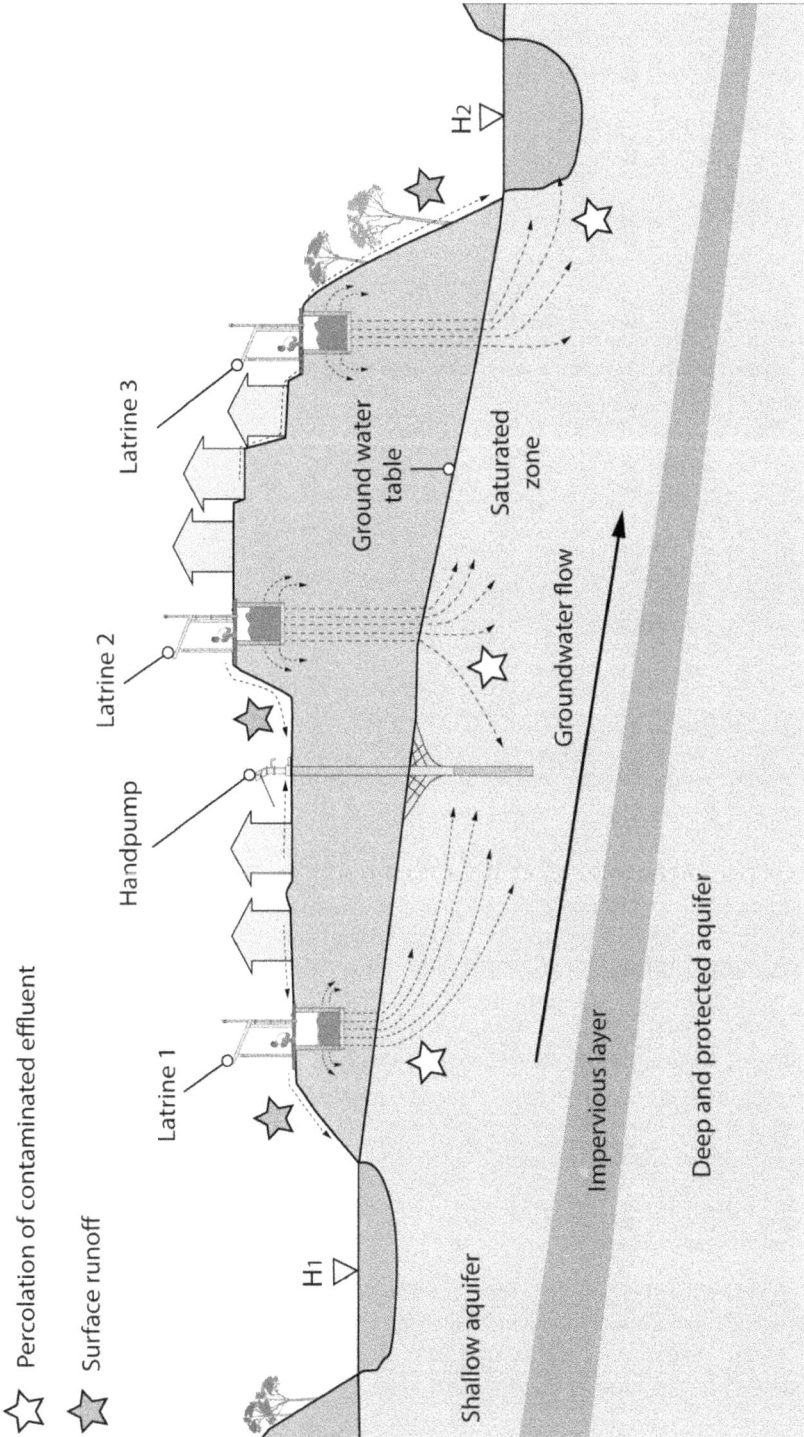

Figure 4.1 Surface and groundwater potential contamination pathways

Source: Adapted from (** Gensch, et al., 2018).

Table 4.4 Minimum separation distances between pollution points and groundwater sources

Soil/rock type	Approximate minimum distance (m)
Silt	10^
Fine silty sand	15
Weathered basement (not fractured)	25
Medium sand	50
Gravel	500
Fractured rocks	Not feasible to use horizontal separation as protection

Note: ^10 m is the minimum distance an infiltration system should be from a water source because of the risk of pollution from localized pollution paths such as fissures, cracks, and disturbances caused by construction.
Source: Adapted from (Macdonald & Pedley, 2001)

The risk of water source contamination is largely based on the travel time of the effluent from the pollution source. This is extremely difficult to measure, even in a non-emergency situation, and is often based on past experience. In the absence of local information, the figures for minimum separation between a pollution source and a water abstraction point are suggested in Table 4.4.

4.4.1 Interpretation

1. Groundwater contamination from toilets must be considered if there is groundwater present below the toilet and it is being used as a drinking water supply (in the present or in the future).
2. Shallow groundwater found in simple wells or shallow boreholes, for example, is nearly always contaminated from other sources, especially in urban areas. It is simpler to change the water source than to explore alternative excreta disposal methods.
3. The pollution load from a single isolated family toilet is minimal (unless it is very close to a water point); problems mainly arise where there is a large number of toilets in a small area.
4. Assessing the likelihood of toilets contaminating a water source is complex and time-consuming. Unless local knowledge exists, use the figures given in Table 4.4.

4.5   Materials and skills

Many of the elements of an FSM intervention are small and relatively simple. It makes strategic sense, particularly during the acute and stabilization phases, to use local materials and skills wherever possible as this will speed up the implementation rate and be easier to operate and maintain. Using local materials and skills has the added advantages of supporting the local economy and providing work, using or developing local skills, and generating income for local people.

The use of specialist equipment requiring particular construction and operational skills may be suitable during the recovery phase, but care must be taken to check that long-term funding and human resources are available locally to ensure sustainability.

4.6   Design life

As an engineer, you will be familiar with designing infrastructure to have a functional life without major rehabilitation; this is also a consideration in emergency relief. There are no agreed figures for design life for emergency infrastructure, although it is very common for it to last a lot longer than originally expected. Table 4.5 suggests approximate design lives for different phases of an emergency; however, a set of design lives agreed by all implementing partners is preferable. What you don't want is different agencies using different design lifespans (e.g. Agency X building toilets for a three-year lifespan and Agency Y building them for a three-month lifespan). In major responses, this will normally be led by the agency responsible for water supply and sanitation coordination, but it could also be part of the role of a sanitation technical working group. In smaller emergencies, it may be up to the agencies delivering the toilets to agree on a common timeframe or to discuss this with local coordinators such as the local government.

In terms of specific technologies, Table 4.6 suggests approximate design lives, although a large safety factor should be added to allow for the rapid and frequent changes that occur during an emergency response. Typically, flash floods and limited storm damage cause days, weeks, or occasionally months of displacement. Earthquakes, drought, and civil unrest generally have a prolonged effect and require both immediate and longer-term solutions.

Table 4.5 Design lives for emergency phases

Emergency phase	Suggested design life
Acute response	6–12 months
Stabilization	1–3 years
Recovery	5–10 years

Table 4.6 Design lives for specific technologies

Emergency phase	Suggested design life
Public toilets for family use	1 year
Toilets for clinics, schools, market areas, etc.	5–6 years
Individual and shared household toilets	5–16 years

4.7 Standardization

Even when responding to quite modest emergencies, it is common for multiple agencies to be supporting the response. Standardization of approach makes the sharing of ideas and equipment easier, produces less confusion with users, and simplifies longer-term operation and maintenance for the authorities left behind.

Decisions on standardization are normally developed by a recognized coordination group such as the WASH Cluster, a technical working group, or the local government.

4.8 Staffing

Emergency FSM programmes are staff-intensive, involving organizational, communication, and monitoring tasks in addition to construction. A family toilet programme, for example, will involve staff in promoting toilets, giving advice on siting and construction, purchasing and distributing materials, producing toilet slabs, and so on. An example of a typical staffing structure for an excreta disposal programme in a camp of 50,000 people is shown in Table 4.7, although staff requirements for off-site FSM activities such as emptying, transport, and treatment/disposal are not included.

Table 4.7 Examples of staff needs for an excreta disposal project for 50,000 people

Phase	Main activities^	Staff type	No.
I Days	Setting up and running defecation fields, hygiene promotion		
	Site management, administration, logistics, etc.	Manager, storekeeper, buyer, administrator, driver	5
	Setting up 10 defecation fields	Labourers and team leaders^^	110
	Maintaining 10 defecation fields	Labourers and team leaders^^	60
	Hygiene promotion and information exchange	Hygiene promoters and team leaders	55
II Weeks	Constructing and operating communal pit toilets		
	Site management, administration, logistics, etc.	Manager, storekeeper, buyer, administrator, driver	8
	Construct communal toilets	Skilled workers, labourers, and team leaders	210
	Cleaning and maintaining communal toilets	Cleaners and team leaders	85
	Hygiene promotion and information exchange	Hygiene promoters and team leaders	55

(Continued)

Table 4.7 Continued

Phase	Main activities^	Staff type	No.
III^^^^ Months	Family pit toilet construction programme		
	Site management, administration, logistics, etc.	Manager, storekeeper, buyer, administrator, driver	8
	Advising population on toilet siting and construction	Toilet technicians and team leaders^^^	55
	Concrete toilet slab production	Masons, labourers, and team leaders	22
	Hygiene promotion and information exchange	Hygiene promoters and team leaders	55

Notes: ^These activities overlap as the situation and programme develops. Defecation fields will be managed while the communal toilets are being built. The family toilets may be started before the communal toilets programme is finished, etc.

^^Workers setting up and then maintaining defecation fields may be the same staff who move straight on to maintenance as soon as the defecation fields go into use.

^^^Toilet technicians may have a role that overlaps with that of the hygiene promoters and it may be possible to combine the two jobs.

^^^^Staff requirements for FSM have not been included but will depend on the system implemented.

Source: Adapted from (Adams, 1999).

4.9   Cost

The availability of funds to support an intervention is usually at its greatest at the beginning of the emergency, reduces dramatically after the first year, and then gradually tails off as time progresses. This tends to encourage investment in infrastructure before the situation has stabilized, leading to inappropriate decisions and wasted resources. It is also why, if it is clear that the population will remain displaced for at least six months, the promotion of household toilets needs to be made at the start. The front loading of funds also tends to ignore the longer-term costs of operation and maintenance, often leading to poor infrastructure sustainability.

CHAPTER 5

The toilet and the environment

5.1 Inclusive design

Any toilet, or block of toilets, must meet the needs of all potential users. This can be achieved through either:

- An individual approach that provides toilet facilities, aids, and equipment for the use of a person or groups of people – this approach is commonly used to meet the current or short-term needs of a family group; or
- An inclusive design approach that aims to create functional environments to accommodate a diverse range of users and can be used equally by everyone irrespective of age, gender, or disability.

In general, new toilets (or toilet blocks) should be designed to be inclusive so that the widest range of users can access and use them safely. They should meet the following principles:

- There should be safe ease of use by as many people as possible without undue effort or special treatment or by segregating specific users.
- There should be freedom of choice and access to mainstream activities, to allow people to participate equally in all activities. Users should be able to choose whether to use a support person or not and whether to use the same or separate facilities.
- Facilities should provide for a range of users' needs and preferences (by age, gender, their social-cultural norms, or ability).
- Facilities should be organized and laid out in a logical or ordered way that is easy for the user to understand.

5.1.1 Location

The best location for a toilet is close to the user's home. Unfortunately, this is rarely possible in the acute phase of an emergency, although individual family toilets should always be the objective. Individual family toilets are what families prefer and they minimize the operation and maintenance inputs required from the managing organization (see Chapter 3, section 6 for more justification). Some nations have regulations regarding toilet provision in emergencies, but this is very rare; providers must assess the local situation to determine what level of coverage is most appropriate. In the absence of local data, the recommendations provided by SPHERE may be used (Table 5.2). Choosing the best location requires consideration of a wide range of issues; these are summarized in Box 5.1. Section adapted from ** Jones & Reed, 2005.

Box 5.1 Factors affecting toilet location

Land tenure: Together with stakeholders, ensure that there is a common understanding of land tenure. Public or private landowners must understand and acknowledge the broad excreta management strategy (e.g. depth of pits, type of cubicle, accessibility for pit emptying, decommissioning measures, etc.).

Accessibility by all users: In general, people will not travel far to use a toilet, (** Sphere Association, 2018) recommends a maximum of 50 m from a dwelling. Some users may have physical disabilities that make walking or seeing difficult, while others may prefer to visit only at night. Therefore, access routes should be designed accordingly (see Table 5.1 for details).

Access for operation and maintenance: All toilets require cleaning and maintenance and some need regular emptying. Good vehicle access is preferable, but at the very minimum pedestrian access is essential.

User safety: This is a key concern for many female and child users. Toilets should not be built in isolated locations that may expose users to attack or harassment. Safety may also include protection from external threats such as flooding and traffic. Access paths should be illuminated and routed through well-used areas. In general, toilet infrastructure should also be illuminated and public facilities guarded, especially at night. Detailed consultation with the user community is essential prior to making decisions on this subject as there may be local circumstances that affect the decision.

Environmental protection: It is essential to protect surface and groundwater sources and food supplies. Where space is limited, toilets should be located downstream from water sources. However, if there is a clash between the location of a toilet and a water source, it is usually better to relocate the water source to an appropriate distance upstream of the toilet (Table 4.4 and Figure 4.1). Users are generally willing to travel further to collect water than to use a toilet!

Privacy: A toilet must provide privacy and dignity for all users, especially for women and girls. From the outset, this can be enabled by careful selection of the toilet's location; it is also provided by the design and layout of the toilet unit.

Impact on others: If the community around the toilet object to its location, it will be vandalized or users will be prevented from entering. Sensitive siting, ensuring that those closest to it are the primary users, normally limits the problem.

Drainage: Toilets placed in areas subject to flooding or in a drainage channel will quickly be damaged or become inaccessible during rain.

Table 5.1 Recommendations for access path**s**

Factor	Guidance
Width	1.8 m recommended, 1.2 m minimum
Gradient	Less than 1:15 recommended, 1:12 maximum
Steps	0.28–0.42 m depth, 0.15–0.17 m height, all steps of same depth and height, handrail always provided
Surfaces	Firm, even, non-slip, and well drained
Cross fall	Kerb and preferably handrail on downhill side of path
Landmarks	Frequent permanent markers along paths to guide visually impaired users
Hazards	Must be clearly marked and fenced

Source: (** Jones & Reed, 2005)

Table 5.2 Minimum numbers of toilet cubicles for various settings

Location	Acute phase	Stabilization and recovery phase
General site provision	1 cubicle for 50 persons (communal)	1 cubicle for 20 persons (shared family)
		1 toilet for 5 persons or 1 family
Market areas	1 cubicle for 50 market stalls	1 cubicle for 20 market stalls
Hospitals/medical centres	Minimum: 4 cubicles in outpatient facilities (separated for men, women, children, and healthcare workers)	Minimum: 4 cubicles in outpatient facilities (separated for men, women, children, and healthcare workers)
	1 cubicle for 20 beds and/or 50 outpatients	1 cubicle for 10 beds and/or 20 outpatients
Feeding centres	1 cubicle for 50 adults	1 cubicle for 20 adults
	1 cubicle for 20 children	1 cubicle for 10 children
Reception/transit centres	1 cubicle for 50 persons	
Schools	1 cubicle for 30 girls	
	1 cubicle for 60 boys	
	1 cubicle for 10 staff – minimum 2	
Offices		1 cubicle for 20 staff
Urinals	1 position for 50 users	

Notes: Cubicles in communal settings should be in the ratio of 3 female cubicles for 1 male, provided that male urinals are also installed.
In communal settings, the same number of bathing cubicles should be provided as toilet cubicles.
The maximum walking distance to a cubicle is 50 metres.
At least 15 per cent of cubicles should be child-friendly and fully accessible to people with mobility difficulties, with a minimum of 1 per communal block.
Source: Adapted from (** Sphere Association, 2018), (Mooijman, 2012), (** Gensch, et al., 2018).

5.2   The cubicle and its surroundings

5.2.1 Accessibility

The best access to a toilet for people with mobility difficulties is a ramp (Figure 5.1); this is also true of access to communal toilet blocks (Figure 5.3). However, where step access is the only option, the installation of wide steps highlighted with bright colours and fitted with handrails is an improvement (Figure 5.4).

A level platform outside the toilet door makes access much easier and is essential for toilets designed for use by those with mobility issues. It is better for doors to open outwards as it improves the space inside the cubicle and is easier to operate for wheelchair users. A handrail on or next to the door is also helpful (Figure 5.5). Figure 5.2 shows the recommended space to leave outside the entrance to an inclusive toilet. In communal toilet blocks such as the one shown in Figure 5.3 the corridor width should comply with the dimensions given in Table 5.1 from the toilet block entrance to the accessible cubicles.

Figure 5.1 Ramp access to a toilet
Source: Adapted from (** Jones & Reed, 2005)

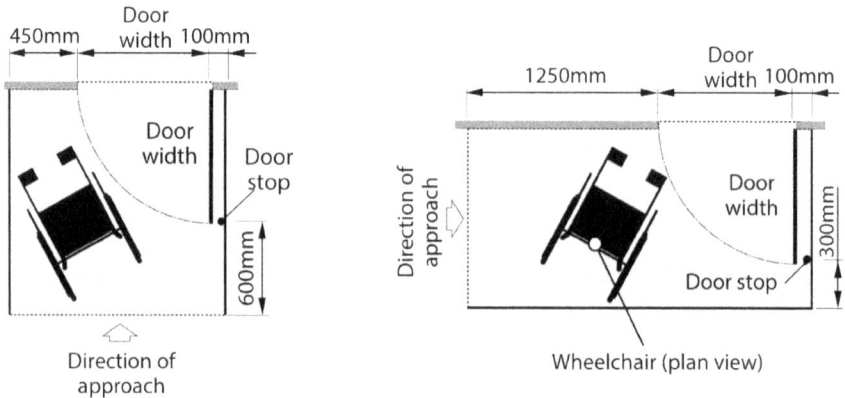

Figure 5.2 Minimum dimensions of the flat platform for opening the toilet door
Source: Adapted from (** Jones & Reed, 2005)

There are advantages and disadvantages relating to the direction of the door opening. Doors opening inwards reduce the effective internal area of the toilet but doors opening outwards obstruct the access passage. Two-way opening doors are easier to use but more difficult to lock (Figure 5.6). See Table 5.3 for further information on door design.

Cubicle for pupils
with special needs
showing handrail

Urinal wall, partially
smooth plastered
against splashes

Air bricks running
complete length of
front and back

Wooden doors

Handrail

Corrugated iron
or plastic roof

BOYS

Door lintels of
suitable material

Backfill around slabs
to seal top of pit

Special needs cubicle door hung on right,
remaining doors hung on left

Handrail

All handwashing area, privacy,
and toilet walls, in concrete blocks

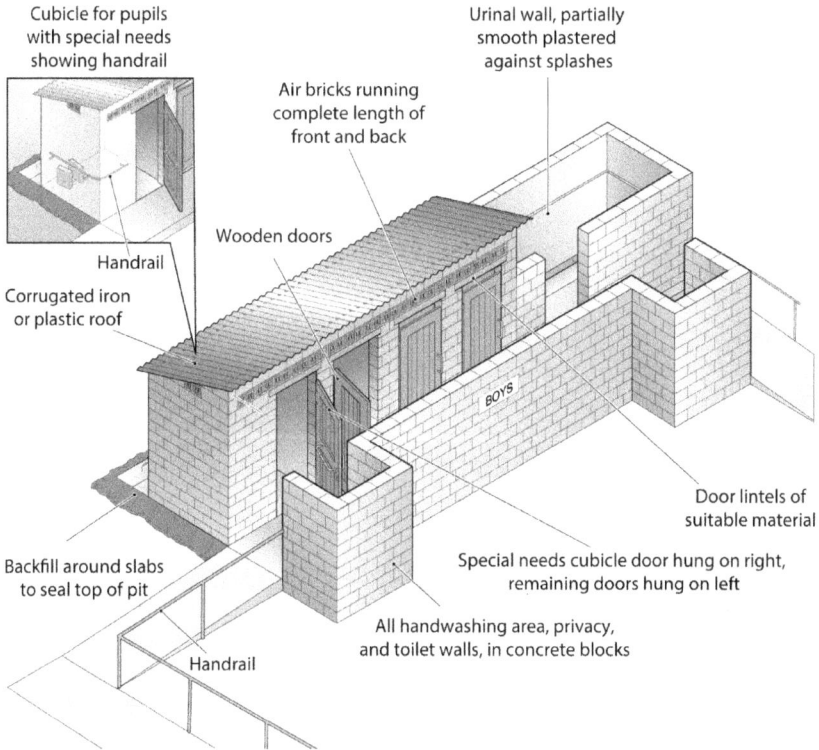

Figure 5.3 Accessible toilet block

Figure 5.4 Accessible step entrance to a toilet
Source: (** Jones & Reed, 2005)

Figure 5.5 Handle extended to full width of door
Source: Adapted from (** Jones & Reed, 2005)

The threshold between the outside and inside of the toilet should be level for easy access (Figure 5.5). If a step is necessary because the toilet must be elevated, then the rise should be marked with a bright colour to assist visually impaired users. If possible, provide additional ramp access.

In communal settings, where not all toilets are designed for people with physical impairments, the door to the toilet cubicle designed for them should be clearly marked with bold signage or a different door colour.

Doors must be high (and low) enough to prevent outsiders looking in and must have a secure door lock that is easy to operate by all users, including those with mobility issues. In some societies, a privacy wall may be needed in front of the cubicle door. This section is largely adapted from **Jones & Reed, 2005.

5.2.2 Toilet cubicle

The main purpose of the cubicle is to provide the user with convenience, privacy, security, comfort, and dignity. The construction design and materials used should generally be similar to those used for the users' homes, as this will facilitate acceptance, ease of acquiring materials, and access to construction skills. Key design considerations are shown in Table 5.3.

Figure 5.6 Large door bolt, easy to grasp
Note: The small window allows the door to be opened from the outside in an emergency.
Source: (** Jones & Reed, 2005)

Table 5.3 Emergency toilet cubicle recommendations

Factor	Guidance
Internal dimensions	Minimum: 0.8 m wide × 1.2 m deep if door opens outwards
	Minimum: 0.8 m wide × 1.6 m deep if door opens inwards
	Ideal: 1.0 m wide × 1.4 m deep if door opens outwards
	Ideal: 1.2 m wide × 2.0 m deep if door opens inwards
	Inclusive design: 2.0 m wide × 2.0 m deep, door opens outwards
Floor	• Top of floor slab should be at least 150 mm above surrounding ground level to prevent surface water entry. • Concrete is ideal, preferably lightly reinforced, cast in situ or prefabricated. Hard-wearing, impermeable, easy to clean. Floors spanning a pit may need engineering design (Table 5.5 and Figure 5.10). There is a wide belief that making concrete is easy, but that is not true: making good-quality concrete is a skilled task (see Box 5.2 for details). • Plastic sheet on compacted base. Suitable for short-term use. • Wood. Satisfactory but difficult to clean and can rot quickly if not treated. • Pre-moulded plastic panels (Figure 5.9). Common in rapid response but unsuitable for longer term. Can be used over a pit if fully supported around the edges. • Compacted earth/mud. Possible for family toilets but unhygienic and poor wear properties for communal facilities. Likely to liquefy/collapse if regularly wetted. • More information on floor slab options in given in Table 5.4.
Walls	• At least 2.0 m high to allow users to stand easily. • Plastic sheeting on a wood frame is acceptable as a temporary measure but is prone to damage, vandalism, and theft and is a security risk for women and girls. Must be firmly fixed to the support frame.

(Continued)

Table 5.3 Continued

Factor	Guidance
	• Any locally available building material that provides the user with convenience, privacy, security, comfort, and dignity.
Roof	• Not always essential (generally depends on the weather) unless specifically required by the excreta disposal technology. • Generally made of locally available materials. • Better to slope towards the rear of the cubicle.
Door	• Width of the door opening generally 0.5 – 0.6 m, but 0.8 m minimum is recommended for inclusive access. • Height depends on the stature of users. Generally around 2.0 m. • Commonly a sturdy design made of wood/plywood with secure locking facilities and preferably self-closing hinges. • Plastic or canvas sheeting can be used as a temporary emergency measure.
Ventilation	• Simple openings in the walls, close to the ground and/or above eye level. • May require netting to keep out rodents and other animals.
Light	• Natural lighting generally an asset, except where excluded by the excreta disposal technology. • Locate above eye height to protect privacy. • Consider artificial light if the communal toilet is used at night.

Table 5.4 Comparison of toilet slabs

Slab type	Comments
NAG magic plastic slab (1.2 × 0.8 m)	Needs no supporting timbers – just ensure that the pit edges are stable. Trench must be no more than 1.0 m wide. Includes foot-operated drop-hole cover. A pour flush defecation point can be inserted into the slab.
Wooden slab	Quick to construct if locally available materials. Difficult to clean. Prone to termite attack and rotting. Can be covered with plastic sheeting to increase life and ease of cleaning. Generally a short-term solution.
Bush timber (logs laid across opening, covered by plastic sheeting and packed earth)	Fast and cheap to construct. Easily upgraded with a SanPlat concrete slab or plastic slab. Difficult to keep clean. Badly affected by rainfall or bathing/laundry activities. Rot or termite damage likely.
Dome slab (1.2–1.5 m diameter)	Good long-term solution. Durable and robust if correctly made. Difficult to transport.
SanPlat slab (0.6 × 0.6 m)	Good for upgrading timber slabs. Quick to manufacture in bulk and at high quality. Hygienic.
Ferrocement slab	Thinner than reinforced concrete slabs but requires training in new manufacturing techniques.
Reinforced concrete slab	Generally a well-understood manufacturing process but important to control quality. Produces hard-wearing and hygienic surface. Heavy.
Plywood slab	Water-resistant plywood expensive and difficult to source.

Source: Adapted from (** Harvey, et al., 2002).

The cubicle design and construction have a direct impact on the health and well-being of users. They also influence whether the toilet is used and looked after. Users, therefore, must be involved in the cubicle design, to ensure that it is socio-culturally acceptable and to encourage users to take responsibility for its management. However, in the acute response stage, it may not be possible to hold discussions with users and so care must be taken to ensure that any cubicles constructed meet the primary aims set out in the previous paragraph. Figure 5.7 illustrates some common designs of cubicles and access options and Figure 5.8 shows possible fixtures that could be added to improve accessibility.

Family toilet

Communal toilets

Figure 5.7 Selection of emergency toilet structures and access

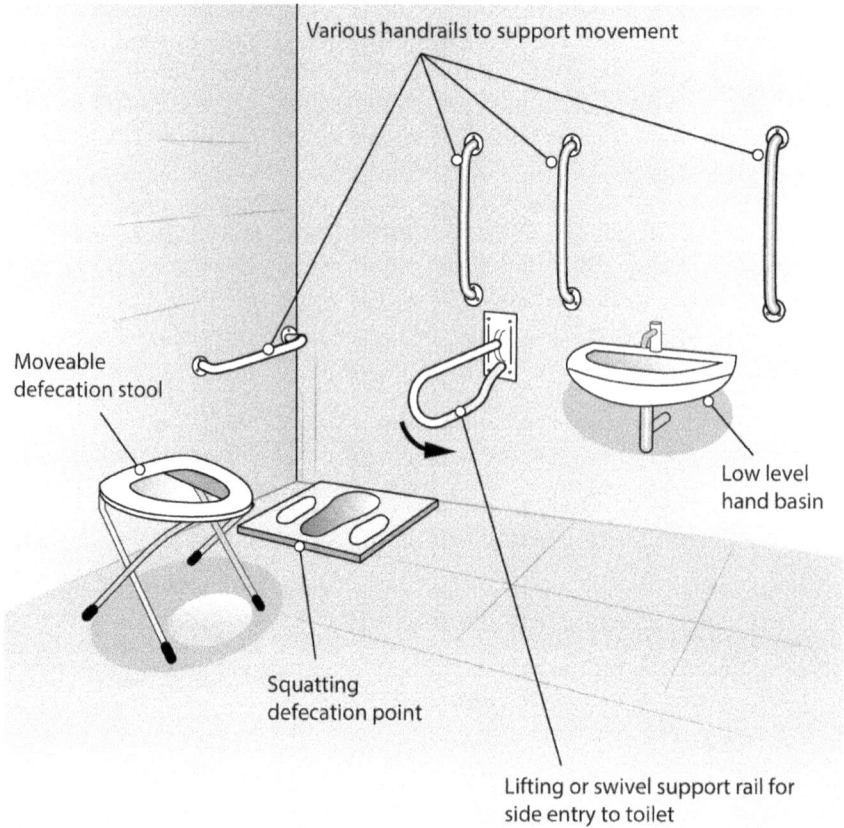

Figure 5.8 Possible fixtures to support toilet accessibility

Box 5.2 Mixing concrete for slab production

Components of concrete

Concrete is made of four ingredients:

- **Cement:** Normal builders' cement is satisfactory provided it is fresh. It should be delivered in its original packaging and contain no hard lumps. It should be stored on-site in a covered area and not allowed to get wet or damp from high humidity.
- **Water:** Clean drinking water is the best but surface water from ponds or streams is adequate provided it is filtered to remove suspended material and organic matter. Saline water is not suitable.
- **Sand:** Commonly known as 'fine aggregate'. It should be clean and free of salt or other impurities. It should be sieved to remove very fine and very large particles, ideally having a variable grain size around 4 mm.
- **Large aggregate:** This should also be clean and free of impurities. It is normally made from crushed natural stone or sieved river gravel. Aggregate made from crushed brick or shells will make the final mix much weaker. The aggregate size should be greater than 9.5 mm and less than one-fifth the total thickness of the slab (around 15 mm).

(Continued)

Box 5.2 Continued

To make 1 m³ of 1:2:4 concrete requires approximately 320 kg cement, 600 kg sand, 1,200 kg aggregate, and 176 litres water.

Mixing

Concrete is usually mixed in the proportion 1 part cement to 2 parts sand to 4 parts large aggregate (1:2:4). Ideally, this should be measured by weight, but generally this is not possible so mixing by volume is appropriate (i.e. 1 bucket of cement to 2 buckets of sand, etc.).

Once the materials are thoroughly mixed together, water is added to start the hardening process. The hardening process is a chemical process and **the less water added the better**: however, if it is too dry it will be difficult to mix everything evenly. Adding too much water will lead to weak concrete. In general, add between 0.4 and 0.6 litres of water to 1.0 kg of cement.

Curing
Once the concrete has been mixed, it will start to cure (harden and gain strength). The time taken for this varies but generally the warmer the air temperature the quicker it will harden. Concrete should not be mixed in freezing temperatures; ice crystals forming in the water will prevent proper curing and reduce the quality of the final concrete. Although curing is a chemical process, it is important to keep the concrete damp and cool for the first 24–48 hours. Cover any exposed concrete surfaces with a moisture-retaining material such as jute or hessian sacking soaked in water. Large flat surfaces can be covered by a shallow layer of water by building a temporary mud dam around the edges of the slab. After seven days, the concrete will be strong enough to move around and transport. After 28 days it will have achieved most of its maximum strength, although it will continue to strengthen indefinitely.

There are many guides to concreting on the internet; for more information, (Cement Concrete & Aggregates Australia, n.d) is suggested.

Table 5.5 Spacing for standard steel reinforcing bars in concrete slabs over a pit

Slab thickness (mm)	Steel bar diameter (mm)	Spacing of steel bars (mm) in each direction for minimum slab spans of:				
		1.0 m	1.25 m	1.5 m	1.75 m	2.0 m
65	6	150	150	125	75	50
	8	250	250	200	150	125
80	6	150	150	150	125	75
	8	250	250	250	200	150

Source: (** Harvey, 2007)

5.2.3 Handwashing

Handwashing with soap is the single most cost-effective measure for reducing the transmission of diarrhoeal diseases. Studies indicate that washing hands with soap and water can reduce the risk of diarrhoeal disease by 42–47 per cent (Curtis, 2003). If diarrhoea is a major problem (and it often is), with evidence or risk of high morbidity or mortality, the focus of response should be excreta disposal, handwashing, protection of water from contamination, and the provision of clean water in adequate quantities. The necessary hygiene education and promotional interventions should similarly focus intensively on these aspects until the risks have been mitigated.

Toilet providers must ensure that all public, communal, shared, and household toilets have handwashing facilities and mechanisms to ensure that they remain functional and usage is enabled for all users (e.g. by taking height into account). During the initial emergency response, the provision of basic handwashing equipment (e.g. jugs of water, soap, and a plastic bowl for grey water collection) is better than delayed provision of improved systems.

Handwashing units come in many different sizes and materials (Figure 5.11), ranging from plastic or metal cisterns with taps to open containers with dippers or bowls. In all cases, the quality of the devices needs to reflect the importance of handwashing with soap in reducing diarrhoeal disease. Improvised or poor-quality solutions risk breakage, getting dirty, or being stolen. In general, the handwashing options illustrated are for the acute phase of an emergency or for family use. Heavy, prolonged use may cause soakaway pits to clog up very fast with sand and soap and produce unpleasant odours. More information on sullage disposal is given in Chapter 14.

The supply of water for handwashing is critical. When there is no piped water supply or point water source (i.e. a well, handpump, or spring) within 30 m of the toilet, provision must be made for water storage and replenishment at the toilet block. During the rainy season, the collection and use of rainwater for handwashing should be encouraged.

Determine the most appropriate type of soap with users. For example, for individual household toilets, users might prefer hard bar soap because it is affordable, long lasting, and easier to protect from theft. Communal and public toilet users might prefer liquid soap or soapy water, which is less prone to being stolen, and is a touch-free dispensing option (unlike hard bar soap).

The selection of culturally appropriate handwashing interventions should be carried out in discussion with representatives of the affected communities. Beneficiary participation is needed to ensure that handwashing facilities and management mechanisms are acceptable and sustainable.

Table 5.6 suggests minimum design criteria for handwashing devices. Toilet providers must establish mechanisms to ensure that handwashing facilities are kept continuously topped up with clean water and soap (before they become empty). In the case of communal toilet blocks, this may require the presence of an attendant. The coverage and condition of handwashing stations must be monitored routinely and problems rectified immediately.

Handwashing facilities have no impact if they are not used. Measures must be taken to actively encourage users to wash their hands after toilet use, after dealing with children's faeces, and before preparing and eating food. The links between excreta and disease must be clearly understood by all. More details on handwashing facilities can be found in (Coultas, 2020).

Table 5.6 Handwashing facilities for public and private toilets

Situation	Provision
Communal/public toilet blocks	1 per 5 toilet cubicles, preferably separate facilities for each sex
Household/shared family toilets	1 per toilet cubicle
Location of handwashing facilities	Less than 10 m from the toilet
Facility design	Accessible by all users, especially children Soap in tamper-proof device Water flow rate of approx. 0.05 litres/second Water quantity of 0.5–1.0 litres/person/day Shallow bowl with plug and drain below water outlet
Water storage	At least half a day's demand
Promotional materials	Footpath from toilet to handwashing station
	Mirrors
	Hygiene promotion posters
Wastewater	Gravel-filled soak pit (see Table 4.3 for appropriate infiltration rates) or connection to an appropriate drainage system (see Chapter 14, section 1 for further details)
Security	Devices must be robust and securely fastened to prevent theft

5.3   The toilet interface

5.3.1 Sitting or squatting?

Of the two, squatting has the most benefits: the human body evolved to defecate squatting, it is the most hygienic as only the bottom of the feet come into contact with the defecation point (assuming the user is not wearing shoes), it is the simplest and cheapest design to provide, and it is the easiest to clean. Having said that, the benefits are of little importance if users refuse to use squatting toilets! Users' views on the correct way to defecate are usually strongly held and difficult to change. The choice must always be based on user preference.

5.3.2 Dry interface

A dry interface is one with the primary purpose of supporting the user while they excrete directly into a receiving chamber. For squatting users, a simple flat surface with a hole is sufficient. Sometimes the floor also has raised areas where users place their feet to prevent them from being contaminated by wastewater, such as shown in Figure 5.9 and Figure 5.10. For users who prefer to sit, a variety of designs are possible, such as those shown in Figure 5.12. These are collectively known as pedestal interfaces. Regular cleaning of the seat is necessary to prevent cross-infection. Dry interfaces often benefit from a tight-fitting lid to reduce fly and odour problems.

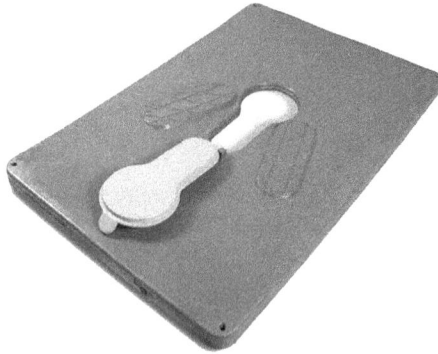

Figure 5.9 Pre-moulded plastic floor slab with defecation hole

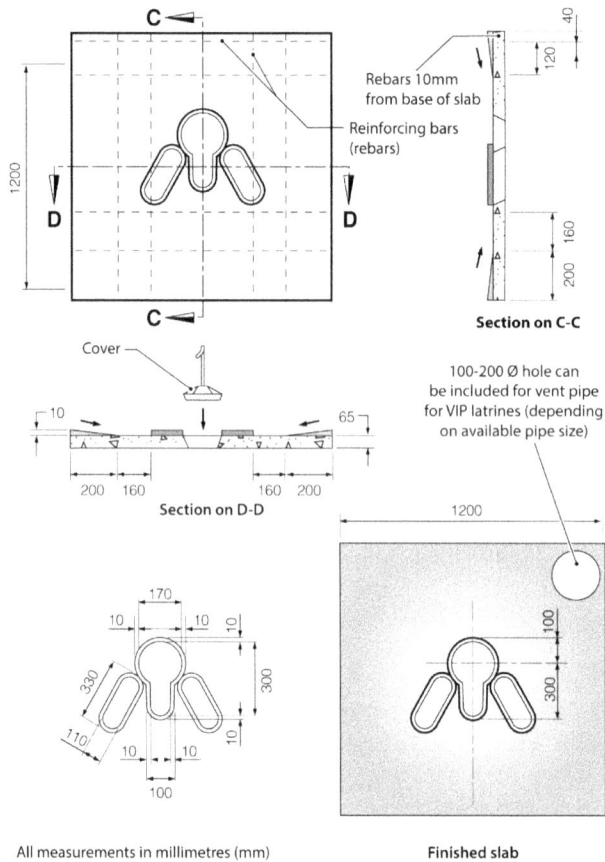

All measurements in millimetres (mm)

Finished slab

Figure 5.10 Reinforced concrete floor slab with adult defecation hole
Note: The size of the defecation hole and footrests can be adjusted to meet the needs of specific users, such as children or the elderly.

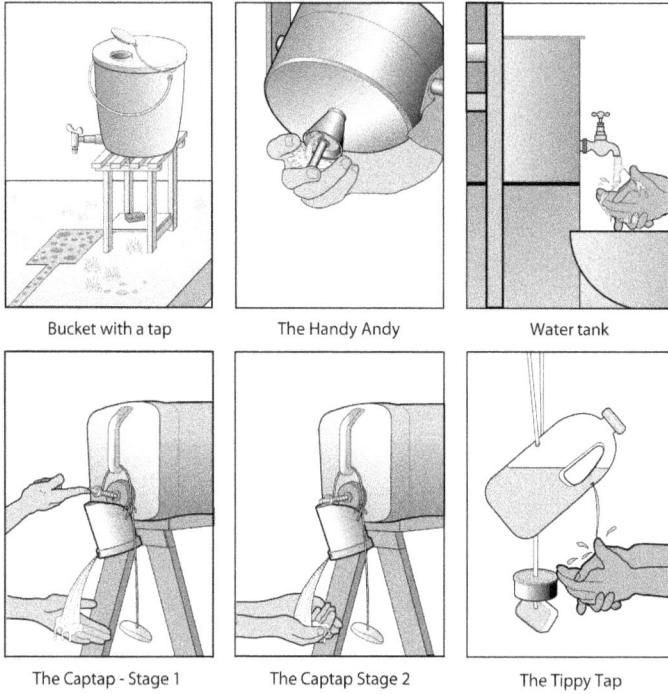

Bucket with a tap | The Handy Andy | Water tank

The Captap - Stage 1 | The Captap Stage 2 | The Tippy Tap

Figure 5.11 Handwashing devices

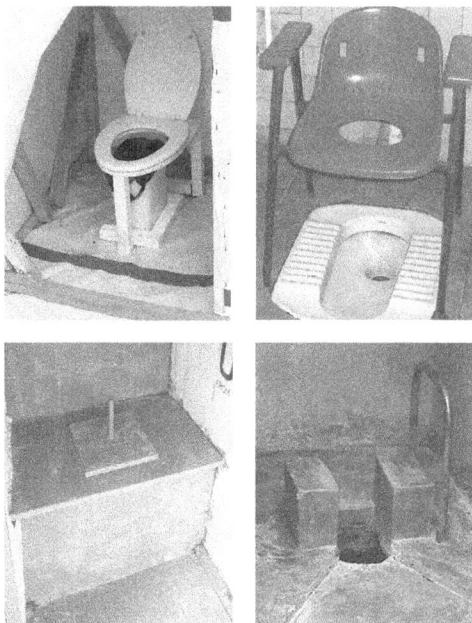

Figure 5.12 Dry pedestal interfaces

Figure 5.13 Urine diversion interface

5.3.3 Urine diversion

Some treatment processes require the separation of urine from faeces, and there is a wide variety of products available to provide this (Figure 5.13). Users who have not seen this type of interface before often find it difficult to use and they must be trained in how to do so prior to provision. They are more commonly used in the stabilization and recovery phases of a response.

5.3.4 Water seal interfaces

Water seal interfaces isolate the toilet cubicle from the rest of the faecal waste management system. This reduces problems with odour and flies in the toilet cubicle, eliminates views of unsightly waste products, gives a feeling of security to users, and can transport the faecal wastes short distances along a pipe. They are the simplest and cheapest upgrade to a plain hole for the defecation point. Water seal interfaces are appropriate only for those who use water or soft toilet tissue for anal cleansing and where the ambient temperature does not regularly drop below zero. There are broadly two types:

- pour flush (Figure 5.14) – require small quantities of water (1–3 litres depending on the length of the subsequent pipe) for direct manual flushing;
- cistern flush (Figure 5.15) – use larger quantities of water (approximately 9 litres) where the subsequent pipe length is longer.

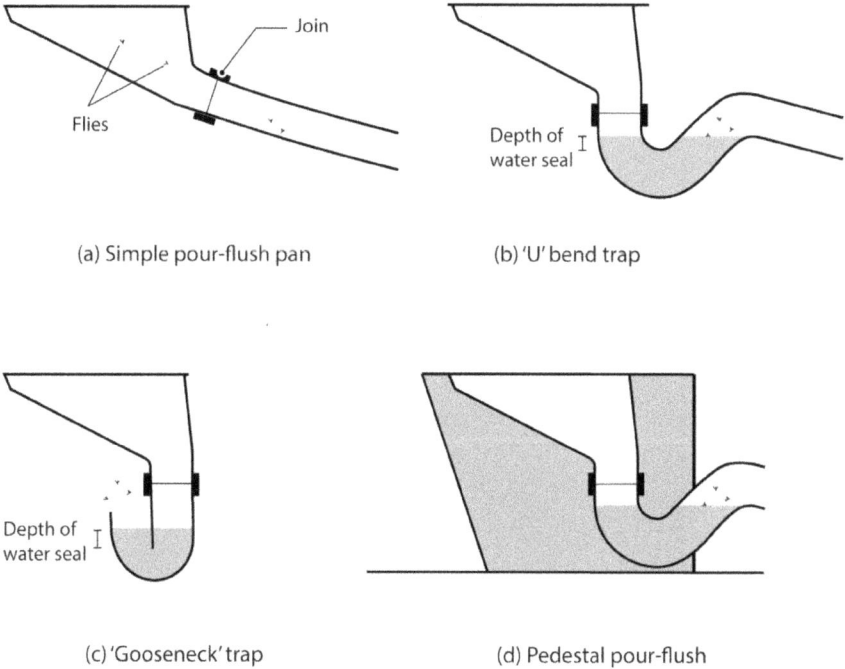

(a) Simple pour-flush pan

(b) 'U' bend trap

(c) 'Gooseneck' trap

(d) Pedestal pour-flush

Figure 5.14 Pour flush configurations

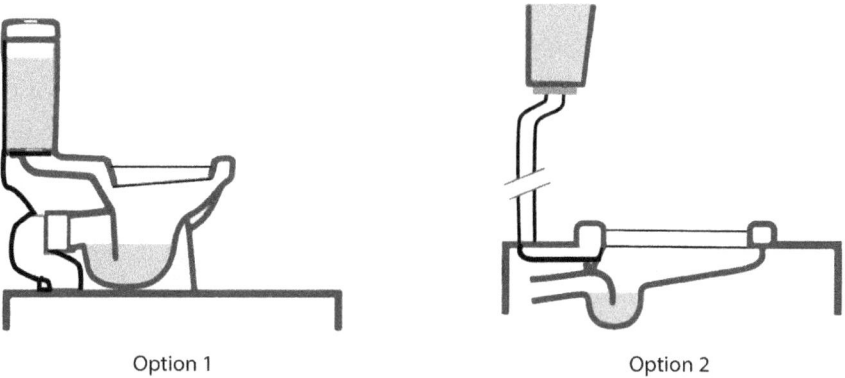

Option 1

Option 2

Figure 5.15 Cistern flush toilets
Source: Adapted from (** Gensch, et al., 2018)

In emergencies, pour flush units are most common as they are cheaper to purchase and easier to install. They are commonly used where exclusively excreta is to be transported through a short pipe.

Figure 5.16 Male urinal designs

5.3.5 Urinals

Male urinals
Providing male urinals reduces the number of male toilet cubicles and the odour and visual unpleasantness of men urinating indiscriminately. Urinals are usually either individual stalls or troughs (Figure 5.16). They may be on the inside or outside of a communal toilet, installed in a household cubicle, or placed separately where men gather. The use of urinals is not common in all societies and it may be necessary to promote them. Urinals commonly discharge to an infiltration pit or a storage tank for treatment and reuse. Men are notoriously inaccurate at urinating so the area around the urinal should be sealed for easy cleaning. Handwashing facilities must be provided. Urinals and their surroundings will produce strong odours if not cleaned regularly.

Female urinals
Female urinals have been manufactured for many years and are becoming common with young people at festivals but they have yet to make an impression in emergencies. Their use could be beneficial because it would considerably reduce demand for toilet cubicles, thus reducing waiting times. They could also be used in homes at night (assuming that they have

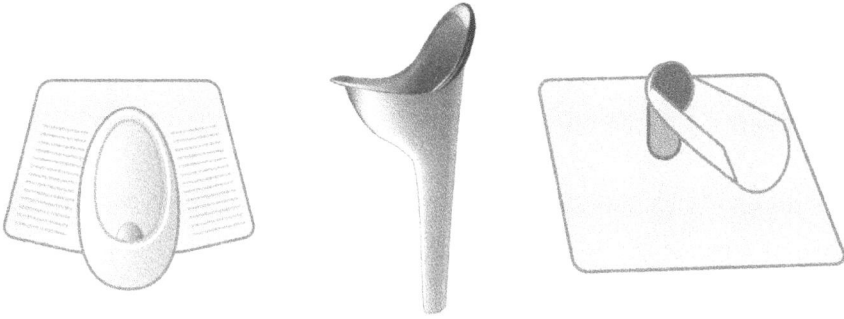

Figure 5.17 Female urinals
Sources: Adapted from (Demiriz, n.d.), (Oxford Devises Ltd, 2023), (Beesley, 2010)

a container to collect the urine safety), which would reduce the security concerns linked to visiting public toilets. To date, there is limited experience of using female urinals in emergency settings and the many social and practical issues to be considered. Extensive discussions with potential users must be undertaken before their use. Female urinals come in two forms: those for individual use and communal units (Figure 5.17). As with male urinals, facilities must be provided for the disposal of the urine, cleaning, and for handwashing.

CHAPTER 6

On-site faecal sludge collection, storage, and treatment

6.1 Immediate actions (first few days/weeks)

In the very early stages of an emergency response, the situation will be chaotic. Populations are still moving around and there will be no effective coordination or management to provide assessments or essential materials. Despite this, it is important to start addressing excreta management. Just because populations are moving around doesn't mean that they will stop defecating. Failure to address excreta management quickly will create a highly polluted environment, exposing vulnerable people to serious diseases.

Anything done at this stage should be deliberately temporary. It is only meant to last a few weeks while the situation partially stabilizes, a better understanding of the situation is gained, and more sustainable infrastructure can be constructed.

6.1.1 Clean it up!

Indiscriminate defecation is not only a health hazard but also unsightly and detrimental to the well-being of an already traumatized community. It is the responsibility of toilet providers to ensure that the environment around the affected populations is free from human faeces. A simple, low-cost method of bringing about both visual and health improvements is to pick up the offending material. Teams of workers in protective clothing and with simple tools can quickly collect the faeces, which can be either buried or disposed of in an existing faecal sludge management (FSM) system.

Design, construction, and operation

Observation walks are a very effective way of identifying and mapping areas with a high defecation load. In some contexts, sanitation brigades may be required at the start of the emergency to clean up excreta. They must be provided with appropriate tools (see Table 6.1) and personal protective equipment (PPE). Any excreta collected must be disposed of safely within a properly designed pit, tank, lagoon, or sewer. Small patches of ground in contact with faeces that have been cleaned should be lightly sprayed with 2 per cent chlorine solution. All tools and PPE used in the clean-up of excreta should be cleaned with detergent and 2 per cent chlorine solution after use. Appendix 2 contains information on how to prepare a chlorine solution.

Table 6.1 Recommended tools for a sanitation brigade of 20 persons

Item	Quantity
Shovels	10
Hoes	5
Rakes	5
Wheelbarrows	5
Gloves (pairs)	20
Boots (pairs)	20
Overalls	20
Backpack sprayers	5
Bleach (2%)	20 litres

Note: Quantities are for guidance only – figures should be adapted to context.

The task should be accompanied by an education campaign to dissuade people from indiscriminate defecation once alternative facilities have been provided.

Advantages
- A quick response to an unsightly and unhealthy hazard.
- Simple to implement using locally available resources and the local population.
- Provides an income to some of the affected population.

Disadvantages
- Should be seen as only a temporary measure until more sustainable facilities can be constructed.
- Close supervision required if it is to be done correctly.

6.1.2 Temporary family toilets

As has already been said (more than once), it is hard to overemphasize the benefits of family toilets (see Chapter 3, section 6). If there is sufficient space and suitable ground conditions, providing each family with its own toilet is the best solution. Initially, this can be something very simple, such as a shallow pit close to where the family is staying (but away from food storage, cooking areas, and water sources). Wooden boards are placed either side of the pit for footrests and a light timber or plastic frame erected around the footrests and clad in canvas or plastic for shelter and privacy (Figure 6.1). Families can be provided with simple tools and basic materials so that they can construct their own toilet. They may need technical advice on what to build and where, and hygiene promotion to maximize the benefits of the toilet. Vulnerable families will need support to construct their toilet.

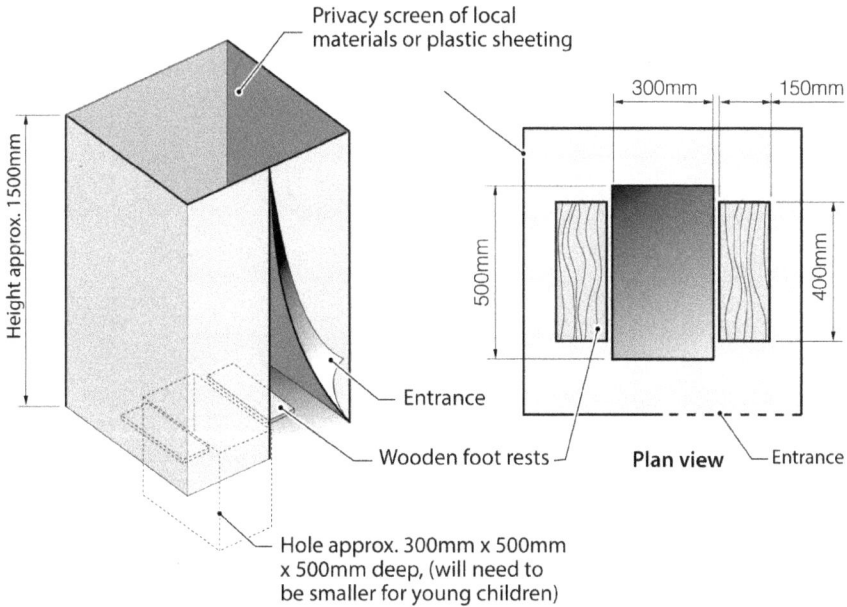

Figure 6.1 Shallow family toilet

After each use, the user throws a small amount of soil on top of the faeces to cover them. This makes the toilet less unsightly for the next user and reduces odour and fly breeding.

Advantages
- Quickly provides families with a private place to defecate.
- Contains excreta below ground to minimize the risk of cross-contamination.
- Can be constructed and maintained by community/family members, with limited input required from external agencies (apart from tools and the basic concept design).
- A simple, cheap design and quick to construct using basic locally available materials.
- Can be moved if the family relocates.

Disadvantages
- Not robust.
- Doesn't control flies, vermin, or odours.
- No protection from the elements.
- Limited security.
- Unsuitable if hard rock or groundwater close to the surface.
- High probability of theft of materials or vandalism.

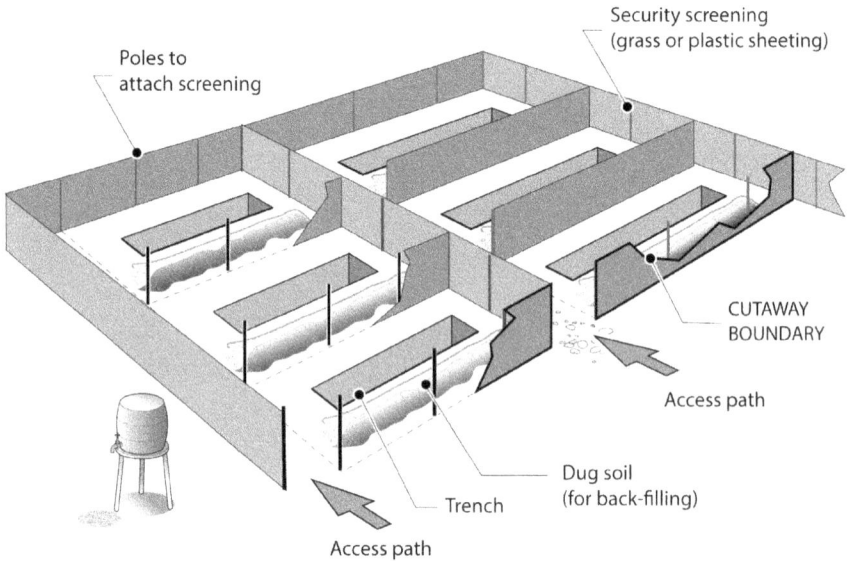

Figure 6.2 Shallow trench toilet

6.1.3 Shallow trench toilets

If temporary family toilets are not possible, shallow trenches can be considered (Figure 6.2). The soil excavated from the trench is left by the side of the trench to allow users to cover their own faeces. This improves hygiene and increases the acceptability of the toilet for future users. Users are encouraged to squat across the trench rather than on one side. Separate trenches should be provided for men and women and possibly small children.

Design, construction, and operation
Trenches are generally 200–300 mm wide and around 150 mm deep. A 150 mm-wide wooden board for a footrest on each side of the trench makes the toilet easier to use and a handrail on one side for users to hold will assist those with mobility difficulties. Small shovels should also be provided for people to place soil on top of their faeces.

Trenches are dug around 1.5 m apart, divided by a privacy screen that is higher than a standing person. In high rainfall areas the trench may have to be covered to prevent it filling with water. An access path is provided between the trenches along the side of each trench on the opposite side to the pile of excavated soil. Provide separate trench systems for men and women. If the users prefer to use water for anal cleansing, provide a water reservoir close to the entrance. Family members would normally provide their own water container, but in some cases it may be necessary to provide these as well. Handwashing facilities should also be provided.

For design purposes, allow approximately 0.25 m² of land per person per day. This is equivalent to 2 hectares (20,000 m²) of land per week for a population of 10,000.

Only a small length of trench should be opened for use at a time to encourage full utilization. Once the whole trench bed has been covered with faeces, it should be closed and filled in.

Advantages
- Rapid implementation (one worker can dig 50 m of trench per day).
- Faeces are immediately covered with soil.
- Employs a large number of low-skilled workers to keep facilities functioning effectively.

Disadvantages
- Limited privacy.
- Uses a large area.
- In hot weather can be highly odorous.

6.1.4 Chemical toilets

In many countries, chemical toilet cubicles have been fabricated for use at festivals and other temporary gatherings. They have various names – Portaloo and Dixi, for example – but all follow a similar pattern: a storage tank containing deodorizing chemicals set below a toilet interface in a cubicle that may also contain a simple hand basin and a flush water tank (Figure 6.3). The storage tank is emptied regularly and the deodorizing chemical recharged each time.

Figure 6.3 Portable toilets, Haiti

Design, construction, and operation
Chemical toilets are prefabricated. They should be conveniently placed for the
expected users and readily accessible to the maintenance vehicle. Toilets are
usually placed in groups with those used by men separated from those used
by women. Alternatively, individual toilets may be provided for a small group
of families. Handwashing facilities are essential if they are not included in the
toilet unit. In some cultures, a privacy screen must be provided. Where the users
do not use water for anal cleaning, toilet paper must be provided. Only excreta,
soft toilet tissue, and handwashing water must be placed in the toilet as other
materials will block the storage tank and cause emptying difficulties.

Advantages
- Rapid deployment.
- Portable and lightweight.
- Easy to install and clean.
- Come in a variety of layouts to suit local customs and practices.
- Usually supplied with a maintenance contract.
- Suitable for short-term use in environments with high water tables,
  floods, hard ground, urban settings, or sites where the landowner has
  not permitted toilet construction.

Disadvantages
- The maintenance contract can make them an expensive option for
  prolonged use (see more on contracts in (** Davis & Lambert, 2002)).
- A secure supply of deodorizing chemicals is essential.
- The provision of handwashing facilities may have to be arranged separately.
- Require very frequent emptying.
- Must be linked to an FSM chain that is not affected by deodorizing
  chemicals.
- Must be located next to vehicular access.
- The site must be perfectly flat to avoid the units falling over.
- Generally designed for users who sit to defecate and use paper for anal
  cleansing.

6.1.5 Bucket toilets

This is a relatively simple way of providing a temporary family toilet, especially
in high-density urban settings, where the ground is hard, or where land tenure
prevents the construction of semi-permanent toilets. As the name suggests,
bucket toilets are locally available watertight containers (around 50 litres),
covered by a plastic toilet seat and cover. Larger containers (up to 200 litres)
are sometimes used for communal settings.

Design, construction, and operation
Buckets should preferably be lined with a plastic bag, as in the pedestal
container shown in Figure 6.4. A tight-fitting cover is preferable to reduce

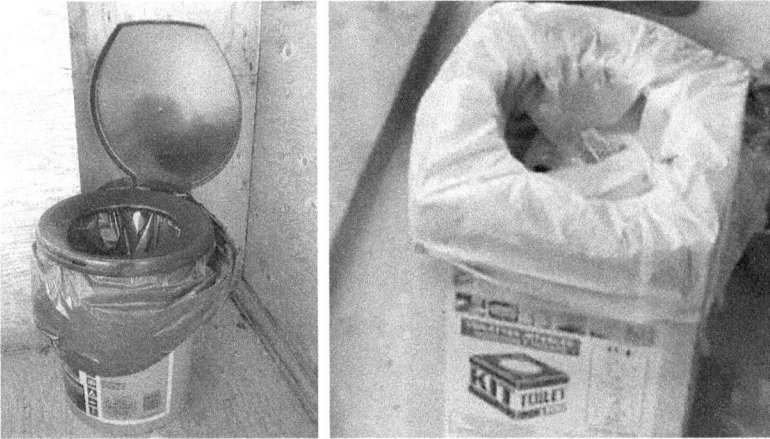

Figure 6.4 Bucket toilets

odour and prevent fly breeding. Where possible, bags should be biodegradable to eliminate the build-up of contaminated non-biodegradable plastic and to prevent the need for bag emptying prior to treatment and disposal. Only as a last resort should non-biodegradable bags be used. Only excreta and soft toilet tissue should be placed in the toilet. Handwashing and anal cleansing water should ideally be disposed of separately (e.g. to a soakaway) as it quickly fills the bag, making it heavy, difficult to carry, and prone to spillage during transport.

Containers such as garbage 'wheelie bins' can be used for communal facilities but tend to fill rapidly, mainly with urine, making them difficult and heavy to handle. The provision of urinals can reduce this problem (see Chapter 5, section 3.5). Container systems can also be used effectively on a larger scale. Rows of garbage bins below the toilet cubicle collect the excreta. When full, they can be removed for processing and replaced with empty, cleaned bins (Figure 6.5). So far, they have mainly been used at festivals, but there could be scope for their use in emergencies.

A regular collection and disposal service is essential to maintain the effectiveness of all container-based systems. When full, bucket toilets can be managed either at the household level (through disposal into a pit or composting system) or at a centralized level. One effective method of managing the collection of bucket toilets is through a swop system. The used plastic bag is removed and replaced with a new, empty one. The top of the used bag is knotted to prevent spillage then taken away in a vehicle with a watertight storage bin for further treatment and disposal. Alternatively, the complete bucket can be exchanged for a cleaned and sterilized empty one. The partially full buckets are loaded onto a pickup truck and taken to a centralized facility for emptying and cleaning.

The management of larger containers is very similar. When full, the bins are removed from below the defecation point, a tight-fitting cover installed, and

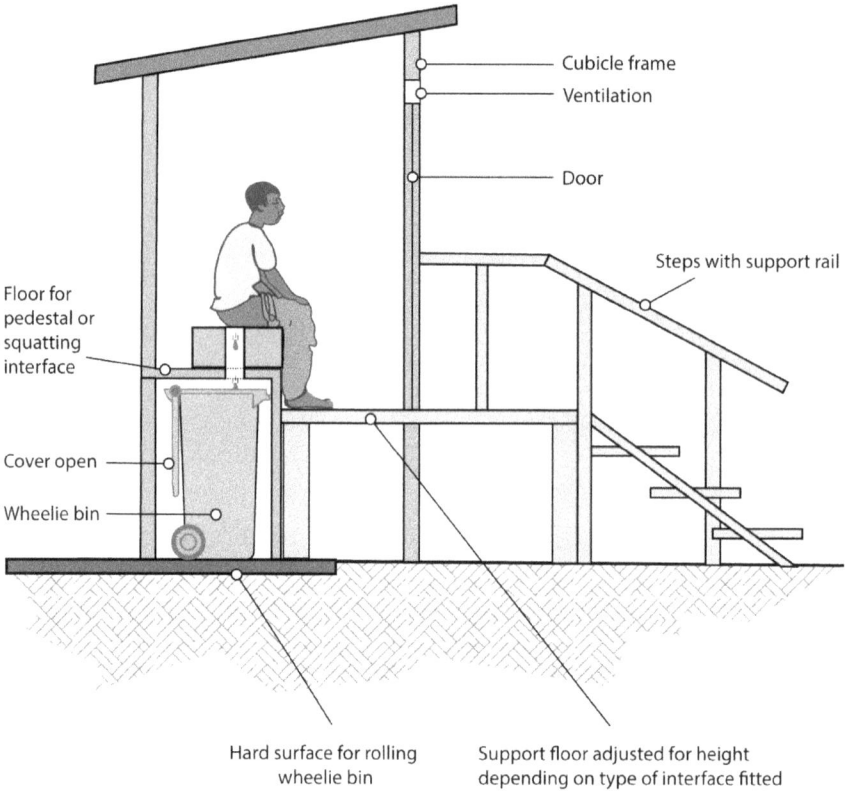

Figure 6.5 Wheelie bin toilet

Box 6.1 Calculating bucket toilet emptying frequency

Approximate emptying period (days) = effective container volume ÷ (daily number of users × waste volume per user)

Where   effective container volume = 75% of total container volume
daily number of users = number of users per cubicle (Table 5.2)
for waste volume per user, see recommendations in Table 4.2

---

the bin transported in a dedicated vehicle to a place for emptying, cleaning, and sterilizing for reuse. A cleaned and sterilized bin replaces the full one as soon as it is removed.

An estimate of the emptying frequency for this and other types of container-based storage systems can be obtained from the calculation in Box 6.1. Note that sufficient vehicles and operators, a place to empty and clean containers, and a site to safely treat or dispose of the waste must all be available.

Advantages
- Quick to assemble and commission.
- If properly managed, does not contaminate the local environment.

- Suitable for short-term use in environments with high water tables, floods, hard ground, urban settings, or sites where the landowner has not permitted toilet construction.

Disadvantages
- Requires systems of safe, well-organized daily collection and disposal.
- Close logistics management essential to maintain a clean and odourless environment.
- Where plastic bags are used for storage, they normally have to be emptied prior to treatment of the waste.

6.1.6 Storage tank toilets

Storage tank toilets use large-volume waterproof tanks (typically 5,000 litre polyethylene tanks) to contain excreta (Figure 6.6). A block of toilet cubicles is constructed on a raised platform with a sewer pipe running below from the toilet interfaces to the storage tank. Water must be added to the system to flush the excreta from the toilet interfaces to the storage tank. This can be added either after each use (like a pour flush pan) or, more often, periodically at the head of the sewer system. Flushing water must be kept to a minimum to prevent it making up a large proportion of the waste collected. A trial and error approach is usually necessary to obtain the optimum amount. Handwashing facilities must be provided. Separate toilet blocks should be constructed for men and women, but they can discharge into a common storage tank.

Only excreta, soft toilet tissue, and handwashing water must be placed in the toilet as other materials will block the sewer or storage tank and cause emptying and further treatment difficulties. The toilets must be closely managed and supervised to ensure cleanliness, frequent flushing, and the

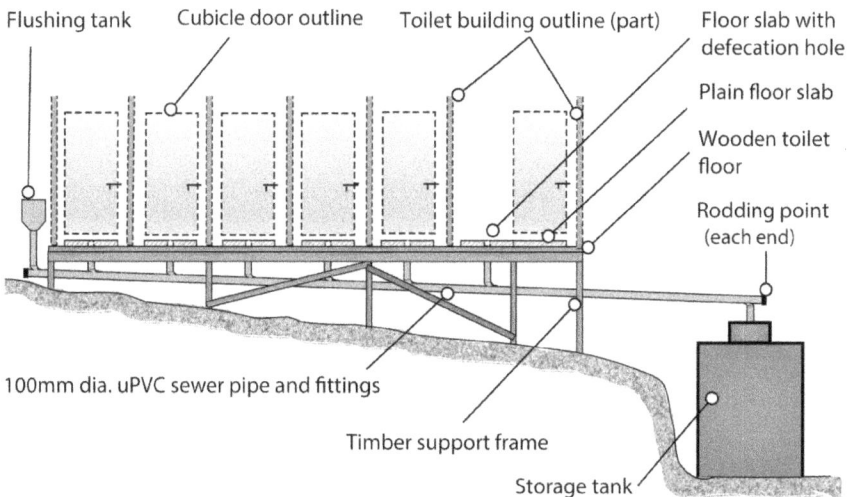

Flushing tank   Cubicle door outline   Toilet building outline (part)   Floor slab with defecation hole

Plain floor slab

Wooden toilet floor

Rodding point (each end)

100mm dia. uPVC sewer pipe and fittings

Timber support frame

Storage tank

Figure 6.6 Communal toilet block connected to a storage tank

safety of users. Emptying is most commonly by a vacuum tanker, but, if the ground slope permits, it may be possible to empty by gravity into a tanker. Storage tank toilets do not provide any treatment, and so the excreta must be taken for appropriate processing.

Advantages
- Can be quickly installed for short-term use in environments with high water tables, floods, hard ground, urban settings, or sites where the landowner has not permitted toilet construction.
- Large HDPE polyethylene tanks are commonly available on the local market.

Disadvantages
- Tanks require frequent desludging, which, if not carried out correctly, may create health risks to the local population, sanitary workers, and the environment.
- Storage tank toilets must be located in areas that are easily accessible by desludging trucks.
- Desludging operations are costly and a clear exit strategy should be sought from the start.
- The connecting sewer frequently blocks from improper use and infrequent flushing.

6.1.7 Adaptation of existing toilets

In many emergencies, toilets will already exist. Often, the quickest response – and the one most acceptable to users – is to renovate them. If toilets are damaged beyond repair, temporary toilets can be constructed, making use of available sewers or other substructure (lined pit, septic tanks, etc.).

Design, construction, and operation
Damaged toilets can be brought into use by replacing toilet interfaces or installing temporary screening to provide privacy. It may also be possible to install additional toilets using elements of an existing sewerage system. As an example, a temporary toilet block could be constructed above a sewer access point, as shown in Figure 6.7. Alternatively, a toilet block similar to that shown in Figure 6.6 could be constructed with the collecting sewer discharging to an existing sewer access point. Always check that the sewer network is functioning downstream of the connection point. When constructing temporary facilities such as those shown in Figure 6.7, include facilities for unblocking pipework and removing gross solids deposited in the toilet.

Advantages
- Users prefer to use facilities they are accustomed to and are more likely to look after them.
- Often quicker than constructing something new.
- Transport, treatment, and disposal facilities are already in place.

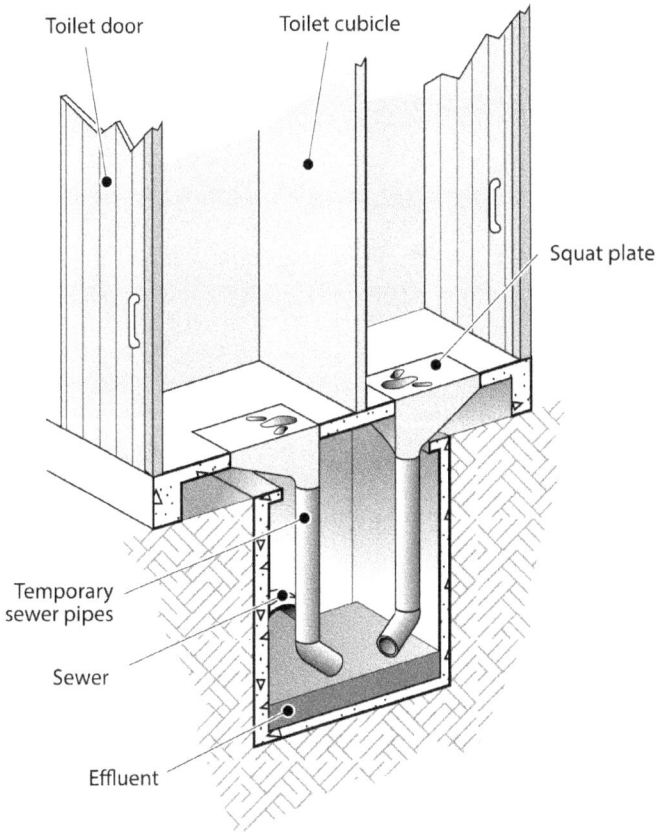

Figure 6.7 Temporary toilet block over existing sewer

Disadvantages
- Pre-existing toilets may have a high water demand to function correctly.
- Only applicable where there is an existing functioning sewer network or containment system.
- Arrangements for cleaning and maintenance are essential.

6.1.8 Defecation fields

If no other alternative is available, in the very early stages of an emergency, the affected population may have to defecate in the open. Defecation fields localize contamination and make it easier to manage the safe disposal of excreta. They have a limited lifespan and can be used only once, so new fields must be prepared in advance of existing fields filling up. Indiscriminate open defecation creates a serious health hazard in affected communities and should be stopped as soon as possible.

Design, construction, and operation

Defecation fields should be made as large as possible. Allow about 0.25 m² per person per day and locate fields according to the distribution of the affected population, to allow easy access for all users. Fields should be at least 30 m from homes, food production and storage and essential institutions such as hospitals. Locate them on land sloping away from buildings, food stores, and surface water sources. Avoid areas liable to flooding. The soil should be easy to dig and to cover the faeces. People may prefer a site with trees and bushes to provide shelter and some privacy, although open land is easier to manage. Ensure that polluted surface water runoff is disposed of safely and does not contaminate downstream water sources. Dividing the field into strips similar to shallow trench toilets, rather than allowing free use of all the field, uses space more efficiently.

Consult on and agree the plans for the defecation fields with users before plans are implemented. This is especially important for people who are not used to defecating in public. Defecation fields need supervision and management, so appoint sanitary assistants to do this job. Designate male and female defecation fields and provide water and soap for handwashing at the exits to the fields. Also provide appropriate anal cleaning material. Illuminate the field and access routes at night.

Encourage users to bury their faeces and anal cleaning materials by providing small shovels to users so that they can dig a shallow hole to defecate into. At the end of each day, sanitary workers should cover all exposed faeces with sand or dry soil to reduce odours and fly breeding. In dry, arid climates, it may be possible to extend the life of a defecation field by employing workers to collect dried faeces at the end of each day and bury them in a pit close to the defecation field, although this is not recommended where users do not wear shoes.

Advantages
- Only a last resort.

Disadvantages
- Become difficult to supervise over time.
- Take up a lot of space.
- Not easy to keep in a hygienically acceptable state.
- Odour, fly, and vermin problems are common.

6.2   Medium-term measures

The toilet elements described in Chapter 5 have been assembled in a wide variety of ways to produce a functioning toilet that meets the needs of users and suits the local environment. This section describes the more common combinations.

IMPORTANT
The primary purpose of a toilet in an emergency is to provide a safe place for users to defecate and to prevent the spread of excreta-related diseases. Sometimes toilets are designed to produce a valuable output, such as cooking gas or fertilizer. This is OK provided it doesn't negatively affect the toilet's primary purpose.

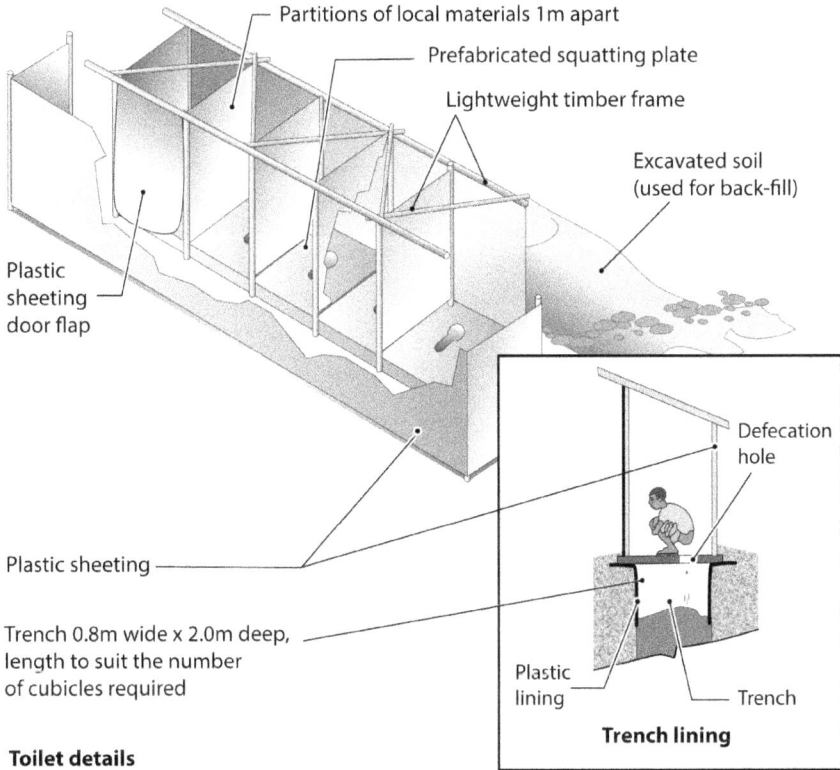

Figure 6.8 Deep trench toilet

6.2.1 Deep trench toilet

Communal deep trench toilets can be quickly prepared to provide a short- to medium-term solution (Figure 6.8). Their lifespan will depend on the number of users and the toilet size, but they generally last around 1–3 months.

Design, construction, and operation
Provide sufficient toilets to cope with peak use in the morning and evening. Design for a maximum of 50 people per metre length of trench (or per cubicle) per day. Adapt usage criteria based on experience, aiming for a more manageable 25 people per cubicle per day. Full construction details are given in (UNHCR, 2015) and (** Harvey, 2007) contains full bills of quantities for a

range of emergency toilet cubicles. Trench toilets are typically constructed in blocks of six cubicles, including at least one for use by persons with mobility issues with a handrail from the toilet entrance to the cubicle. This cubicle can also be used by children if accompanied by a guardian. Alternatively, a separate children's toilet can be constructed that meets their specific needs.

Communal trench toilets must be constantly supervised and maintained if they are to remain in a sanitary condition. Toilet supervisors must regularly clean the floors, defecation points, and surrounding area, and periodically (preferably daily) cover the trench contents with 50–100 mm of soil, raked level. The success of communal toilets relies heavily on the cooperation of the users. Therefore, community representatives must be involved from the beginning. Provide anal cleansing materials, soil for covering excreta, and a handwashing station with water and soap.

Excavation in collapsing soils is one of the most dangerous construction activities; Appendix 3 gives more advice on how to do it safely. Protect the edge of the trench from erosion and soiling with faeces by installing a 1.5 m-wide sheet of plastic or oiled cloth attached to the underside of the floor and hanging down the side of the trench walls. Table 6.2 suggests the soil types that require temporary lining. Arrange for about 0.5 m of the sheet to hang inside the trench. Use local materials or plastic sheeting for the partitions. Dry excavated soil should be given to users as they enter so that they can cover their faeces when finished.

Table 6.2 Lining requirements for different soil types

Soils that DO required lining	Soils that PROBABLY DON'T require lining
Soft sands and gravels	Soils with significant clay content (except where subject to heavy rainfall or rising water table)
Unconsolidated soils	Most consolidated sedimentary rocks
Filled land	Soils with high proportion of iron oxides (laterites)
Compressed mudstones and shales	
Wet clay	

Source: ** Harvey, 2007.

In areas subject to rainfall, cover the site during excavation and place the spoil on the uphill side of the trench. Ensure that the site is fenced off and covered at all times when construction staff are absent to prevent animals or children from falling in. Dig cut-off ditches to divert surface water. Form an overhanging roof with plastic sheeting to divert rainwater away from the trench to prevent it collapsing. Cover the excavated soil to prevent rainfall erosion. Design the cubicle for ease of dismantling and re-erection over a new trench. If required by the users, erect a privacy screen of local materials or plastic sheeting in front of the entrance to the cubicles.

Decommission the trench when the contents reach 0.5 m from the surface. Remove the cubicle, fold the plastic ground sheet over the trench contents

Figure 6.9 Construction of a pedestal trench toilet

and backfill with the remaining soil. Leave the backfill mounded to allow for settlement.

Advantages
- In stable soils, can be installed very quickly using mechanical excavators provided the floors and cubicles have been fabricated in advance.
- With mechanical excavators, they are a great 'day 1' solution that can be upgraded several weeks later.

Disadvantages
- Complex and dangerous to construct in collapsing soils and areas with a high water table.
- Problems with odour if trench walls are fouled.
- Toilets with pedestal defecation points more complex to construct (Figure 6.9).

6.2.2 Simple pit toilets

Many communities in both peri-urban and rural areas rely on pit toilets such as the one shown in Figure 6.10 for disposing of their excreta. They also make an excellent solution during an emergency response. They consist of a simple toilet cubicle with a squatting or a pedestal defecation point situated either directly over or adjacent to a pit. Faeces and urine are deposited directly in the pit where consolidation, dewatering, and biological decomposition gradually reduce their volume. Traditionally, pit toilets were designed to be used until

Air vent

Superstructure designed
and built with appropriate
local materials

Floor slab of wood or
concrete at least 0.15m
above ground level with hole,
covered when not in use

Tight-fitting lid

Pit

Foot rest

Mound of excavated soil
to seal pit lining and prevent
flooding of the pit by
surface water

Perforated lining to
allow liquids to percolate
into the soil

Gases escape into
surrounding soil

The pit should be at least
2.0m deep and 1.0 to 1.5m
preferably round

Solid residue
decomposes and
accumulates

The bottom of the pit should
be at least 1.5m above the
water table especially where
groundwater is used for
water supplies

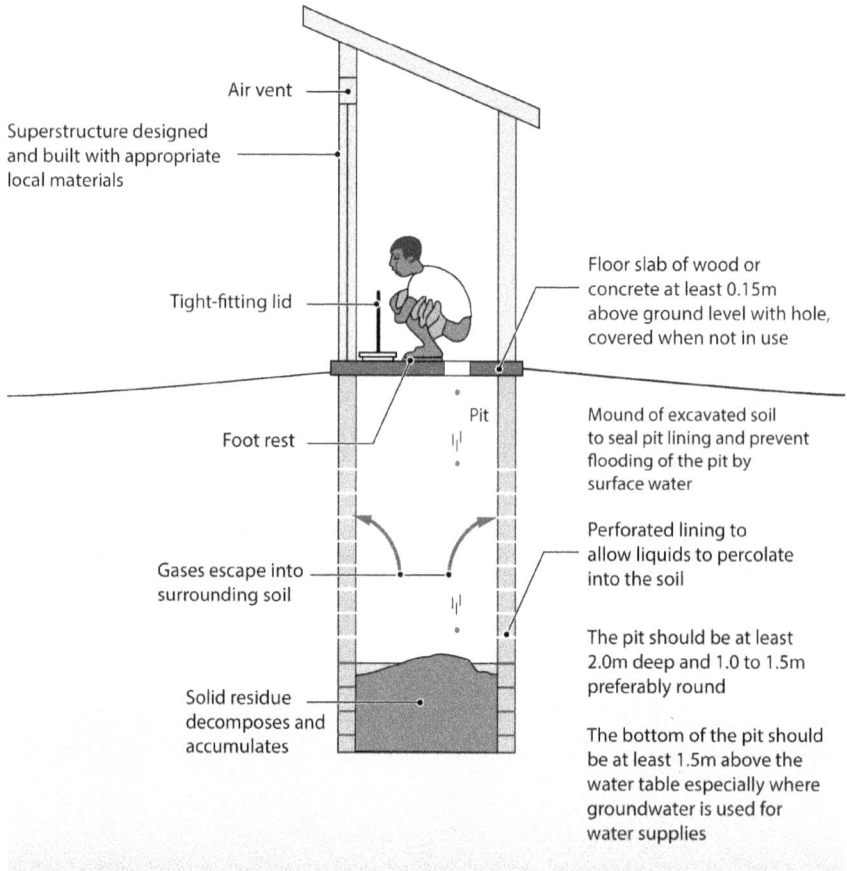

Figure 6.10 Elements of a pit toilet

the pit was full then replaced by another pit, but it is now more common
(especially in urban areas) for pits to be emptied and reused. Prefabricated
toilet cubicles complete with a floor and defecation point can speed up
construction and make transfer between a full and empty pit easier.

Design, construction, and operation
A number of factors must be considered regarding their design and construction.
The design of the cubicle, floor and defecation point have already been
discussed in Chapter 5, section 2.2, so this section will focus on the pit.

Most single pits for household or family use are about 1.0 m across and
3.0 m deep, although in emergencies a shallower pit may be satisfactory as a
temporary solution. Remember that the toilet floor should be at least 0.15 m
above the surrounding ground level to reduce the risk of flooding. This may
require the top of the pit to be built up slightly. It is difficult to excavate pits
less than 0.9 m across because there is not enough room for the person to
work. There is no maximum size for a pit and sizes vary greatly, but pits over

1.5 m wide are expensive and difficult to cover. A worked example of a pit design is given in Box 6.3.

The best shape for a pit (in plan view) is circular. Circular pits are more stable because of the natural arching effect of the ground around the hole and there are no sharp corners to concentrate the stresses. Square and rectangular pits are much more likely to need supporting and require a bigger area of lining than a circular pit of the same internal volume. However, many communities prefer to excavate square or rectangular pits as their construction is similar to the process used for building domestic houses. The size of pit required depends on the number of users, time between emptying or until full, anal cleansing materials used, and ground conditions (levels of permeability). See Table 4.2 for more details.

The top 0.5 m of a pit should always be lined, but the decision as to whether to line the rest of the pit will depend on the type of soil in which the pit is dug. When a pit is first excavated, the walls may be self-supporting, but they may collapse at a later date. This is particularly true in areas with pronounced wet and dry seasons where a pit wall dug in the dry season may collapse in the wet season. One way in which this can be assessed is to examine other excavations (such as hand-dug wells) in the area. If existing excavations have not collapsed and are not lined, then it is fairly safe to assume that pit toilet excavations will not need lining. Where there is doubt, it is advisable to line the pit. Table 6.2 suggests the types of soil that, in general, do and do not require lining. However, all pits that are intended to be emptied for reuse must be lined. Pits can be lined in a variety of materials, as shown in Box 6.2. The bottom of pits should not be lined and the pit walls should be porous except for the top 0.5 m. In most cases, it is not necessary to ventilate a pit toilet.

The safety of workers excavating pits is likely to be your responsibility. Provide adequate ventilation and protective clothing and guard their safety. Unstable soils must be shuttered prior to lining (see Figure A3.1).

Box 6.2 Commonly used pit lining materials

- **Wood:** Time-consuming and difficult to position cross-struts to provide a proper retaining wall. Prone to rotting even when treated.
- **Concrete blocks:** Can be built honeycomb style to allow good infiltration provided the gaps between blocks are limited to the standard width of a mortar joint. Circular block moulds can be used for circular pits.
- **Bricks/stone:** Time-consuming but may be a preferred alternative to concrete blocks if locally available. In very cold climates bricks can absorb water and shatter.
- **Mud/cement stabilized blocks:** Local alternative to concrete blocks or bricks. The addition of cement improves strength and durability.
- **Pre-cast concrete rings:** Liquid cannot escape easily unless the ring is made with drainage holes. Ring moulds required and expensive.
- **In-situ cast concrete:** Relatively time-consuming and requiring skilled labour. No infiltration, therefore pits must be emptied. Expensive.
- **Sandbags:** Sand and bags usually locally available and low cost. Cement can be added to the sand to increase strength and durability. Generally not a long-term solution.

(Continued)

Box 6.2 Continued

- **Oil drums:** Holes must be made in the sides for liquid to infiltrate. Small diameter limits diameter of pit size and ease of excavation. Not a long-term solution as corrode easily.
- **Ferro cement:** Time-consuming and relatively expensive. Works well for supporting flat walls. Impermeable unless holes specifically added during construction.
- **Corrugated-iron sheets:** Very little infiltration can take place unless holes made. Need support bracing.
- **Tyres:** Large quantity required. Allow infiltration through spaces and provide stability but small volume leads to rapid filling. Difficult to empty.
- **Bamboo/cane:** Rots faster than wood and less strong, but may be in more plentiful supply in some areas and encourages community participation and income generation.

Source: **Harvey, 2007.

Box 6.3 Design of a simple pit

Assume: A pit is required to serve four families. Average family size is five and expected time between emptying each pit is six months. Families use paper for anal cleaning and a small quantity of water for handwashing. The soil is porous and the water table is more than 10 m below the surface.

Total number of people defecating in the pit = $4 \times 5 = 20$. However, 5 is only the average family size – some families will be larger and some smaller.

Assume a maximum number of users of 25.

Using Table 4.2, the amount of waste deposited in the pit each day by one person is:

Faeces – 0.25 litres

Urine – 1.4 litres

Anal cleaning paper – 0.002 litres (assume use of newspaper that won't biodegrade quickly)

Handwashing water – 2.0 litres

Total = 3.652 litres

Most liquids will soak away through the pit soil the effective volume is $3.652 - ((1.4 + 2.0) \times 0.75) = 1.102$ litres

Over a six-month period there will be some consolidation and biodigestion of sludge, say 30 per cent

Total sludge accumulation over a six-month period = $1.102 \times 0.7 \times 25 \times 365/2$
$$= 3,519.5 \text{ litres} = 3.52 \text{ m}^3$$

Assume a pit diameter of 2.0 m. This is larger than normal but allows for siting four toilet defecation points directly above the pit.

Sludge thickness in the pit after six months = $4 \times 3.52/\pi \times 2.0^2 = 1.12$ m

Allow for 0.5 m above maximum sludge level for fly and odour reduction plus allowance for variations in usage and desludging times – gives a total pit depth of 1.62 m

Say 1.6 m

In practice, this would be an initial design figure that would be adjusted in the light of user experience.

Construction cost and time

Sufficient allowance must be made at the planning and budgeting stage for the significant cost of pit construction, especially if full lining is necessary in collapsing soils.

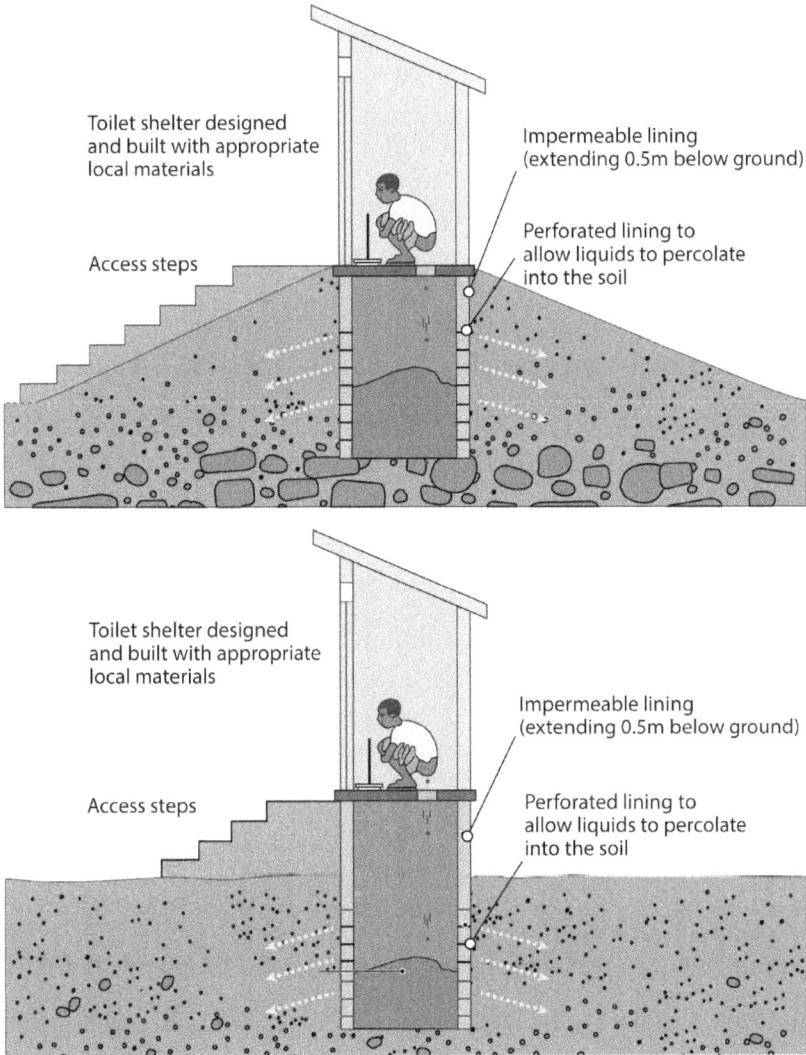

Toilet shelter designed
and built with appropriate
local materials

Impermeable lining
(extending 0.5m below ground)

Perforated lining to
allow liquids to percolate
into the soil

Access steps

Toilet shelter designed
and built with appropriate
local materials

Impermeable lining
(extending 0.5m below ground)

Access steps

Perforated lining to
allow liquids to percolate
into the soil

Figure 6.11 Raised pits

Toilets in rocky or shallow groundwater areas: In these situations, it may only
be possible to dig a shallow pit. If necessary, extend the lining above ground
level to create a sealed chamber for increased storage capacity (Figure 6.11).
Make sure the toilet is accessible by all expected users, especially those with
mobility difficulties.

Toilets in low permeability soils: The quantity of water entering the pit must
be kept to a minimum. Adding water from laundry, bathing, or cooking uses
should be prevented. A small amount of water, such as from urine and anal

cleansing, helps decomposition, but excessive amounts can make the contents offensive, provide breeding grounds for mosquitoes and flies, and possibly result in flooding of the pit. Ensure that surface runoff is directed away from the pit. This can be done by using earth dug from the hole to raise the floor of the toilet above the surrounding land. In some climates the provision of a ventilation shaft can help to evaporate liquids from the pit.

Advantages
- In-situ deep burial of excreta is a safe and proper option for FSM.
- Simple design, constructed using local materials and skills.
- Can be adapted to suit local traditions and customs.
- Effectively removes excreta from the human environment, helping to prevent the spread of disease.
- A widely used technology that is accepted and understood by users.

Disadvantages
- Can give off offensive odours and be a source of fly and mosquito breeding – tightly covering the defecation point when not in use can reduce this problem.
- Digging and lining a deep pit, especially in collapsing soil, can be dangerous, very expensive, and complex.
- Large numbers of pits in a small area could contaminate groundwater (see Chapter 4, section 4).
- Their long history of being poorly designed, constructed, and maintained can lead users to not accept them.

6.2.3 Adaptations to simple pit toilets incorporating the water seal

The simple pit toilet is a good solution, but it does have some disadvantages (see earlier in this chapter). These can often be overcome by the inclusion of a water seal interface (see Chapter 5, section 3.4). There are a number of options, each with its own advantages and disadvantages, but all are feasible only for communities that use water or soft toilet tissue and water for anal cleaning. The options are summarised in Table 6.3.

6.2.4 Ventilated improved pit toilet

For the many communities that don't use water for anal cleansing, the alternative improvement is the ventilated improved pit (VIP) toilet (Figure 6.15). Alterations to the design of the toilet cubicle and the ventilation of the pit can, if done properly, bring about a marked reduction in odour and fly infestations.

Design, construction, and operation
VIP toilet design is very specific and failure to follow the design completely will lead to the toilet not functioning as expected.

- The cubicle must be of solid construction and kept dark. No windows are allowed and the door should be self-closing. A small amount of light is acceptable so that users can see what they are doing and for ventilation purposes. If possible, the door or ventilation holes should face towards the prevailing wind.
- The ventilation pipe should be connected directly to the pit. It must be straight, at least 150 mm in diameter, and extend at least 0.5 m above the top of the toilet roof. The top of the pipe must be covered with fly or mosquito netting.
- The defecation point can be either a pedestal or squatting type, but it must never be fully sealed. This is so that air can pass through even if a cover is in place.
- Natural air flow produces a negative pressure at the top of the ventilation pipe that draws air up from the pit below. This is replaced by fresh air that enters through the toilet cubicle and down the defecation point.
- Any flies that gain entrance to the pit are drawn to a light source as a means of escape. Since the toilet cubicle is dark, the only light source is the top of the ventilation pipe. However, they cannot exit through there to spread disease because the fly mesh over the top of the pipe prevents them leaving.

The complexity of the toilet design and the significant disadvantages it has (see below) suggest that recommending the use of VIP toilets should only follow a serious and extended evaluation of other options.

Advantages
- A successful method of controlling odour and flies in a pit toilet for those who use solid materials for anal cleaning.
- Appealing to some users as seen as 'modern' in comparison to previous toilet facilities.

Disadvantages
- Very expensive compared with other pit toilets.
- Takes longer to construct.
- Most properly designed VIP toilets are built as part of projects – people who copy the design tend to reduce the size of the vent pipe to cut costs and because they don't understand its purpose and they also construct a light cubicle because they don't like a dark one, but both changes defeat the primary functioning of a VIP toilet.
- Many users don't like a dark cubicle, children may be frightened of entering, and a dark cubicle can be a barrier to use by many with mobility issues.
- Sometimes, on cool still mornings, the air flow can reverse, filling the cubicle with strong odours.
- As with all deep pits, they are difficult to dig in areas with a high water table or rock close to the surface.
- There are far more examples of VIP toilets that don't work as planned than of those that do.

Table 6.3 Pit toilets with water seal interfaces

Name	Design, construction, and operation	Advantages	Disadvantages
Pour flush toilet (Figure 6.12)	After use, a small quantity of water is thrown forcibly into the pan, washing the excreta into the pit below and replenishing the water seal.	• A simple low-cost adaptation that prevents odours, flies, and mosquitoes from leaving the pit. • Hides a view of the pit contents, adding confidence to users. • Where soils are porous, the toilet can be used for bathing and laundry. • A widely used technology in Asia. • Easy for others to copy.	• Requires 1–2 litres of water (doesn't need to be clean) to flush the pan.
Offset pour flush toilet (Figure 6.13)	The connecting pipe should be less than 2 m long with a slope not less than 1:30. The pipe diameter is governed by the outlet diameter of the pour flush pan but is generally 75–100 mm.	• The pit is readily accessible for emptying. • The toilet cubicle can be located within the home, improving privacy and security. • When the pit is full, a second one can be dug next to the existing one, the connecting pipe re-routed, and the toilet can continue to be used.	• Slightly more expensive because of the need for a connecting pipe and two floors. • Requires around 5 litres of water to flush. • Unsuitable in highly impermeable soils because of liquid build-up in the pit.
Twin pit pour flush toilets (Figure 6.14)	The discharge pipe is connected to the pour flush pan and enters a small access chamber where it divides into two pipes, one going to each of the two pits. Each pit should have a sludge storage capacity below the inlet of at least one year (two years is preferable). A pit 1.0 m deep and 1.0 m in diameter will last most families over two years and still leave a 0.5 m gap between the sludge and the top of the pit. Pits are filled alternately by blocking opposite discharge pipes. The second pit is used only when the first is full. When the second is full, the contents of the first pit will be safe to handle and it can be emptied and reused.	• There is no handling of fresh excreta during emptying. • Sludge can be safely handled and disposed of locally. • Users like the idea of a toilet lasting 'forever'. • It is often cheaper to dig two shallow pits than one deep one. It is also safer. • Can be used in areas with high groundwater levels or rock close to the surface.	

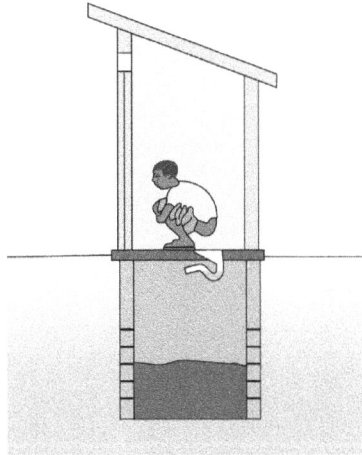

Figure 6.12 Simple pour flush toilet

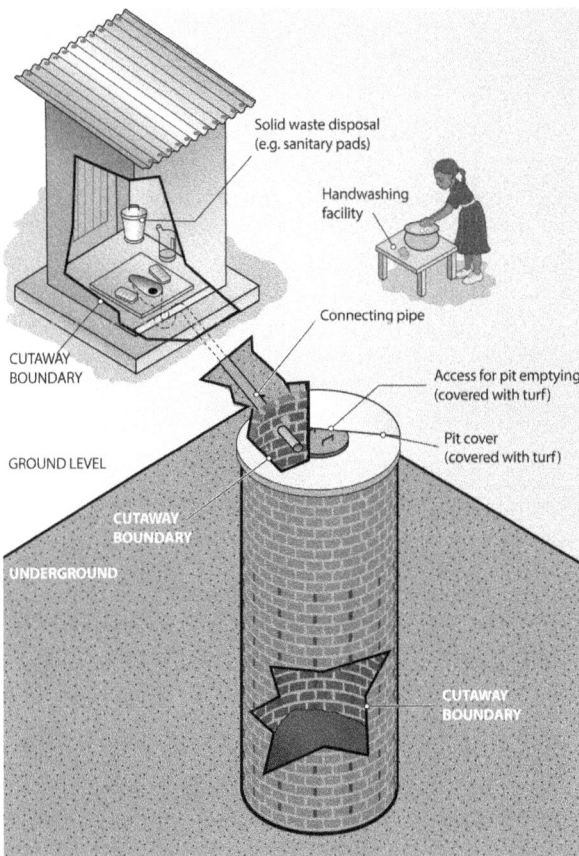

Solid waste disposal
(e.g. sanitary pads)

Handwashing
facility

Connecting pipe

CUTAWAY
BOUNDARY

Access for pit emptying
(covered with turf)

Pit cover
(covered with turf)

GROUND LEVEL

CUTAWAY
BOUNDARY

UNDERGROUND

CUTAWAY
BOUNDARY

Figure 6.13 Offset pour flush toilet

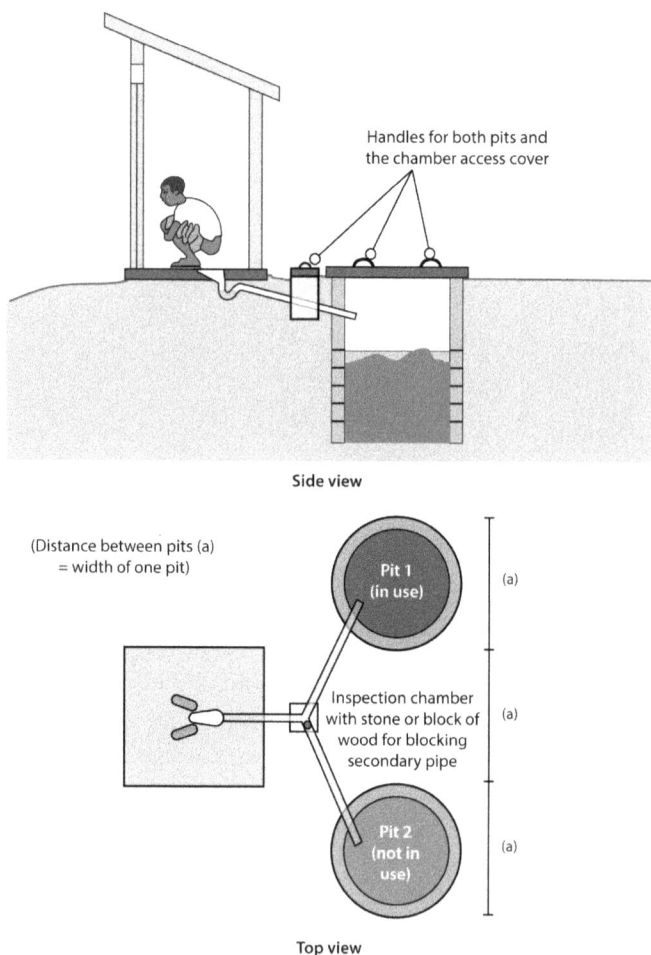

Figure 6.14 Twin pit pour flush toilet

6.2.5 Twin pit direct entry toilets

These consist of two pits (with or without a ventilation pipe) and a single cubicle that either covers both pit defecation holes or that can be moved from one pit top to the other (Figure 6.16). As with the twin pit pour flush toilet, the pits are used alternately, allowing one to fill while the other decomposes the wastes from a previous filling. This is not a common technology in emergencies.

Design, construction, and operation
For a family toilet, each pit has a volume of about 1.0 m³, enough to take between one and two years to fill. The two pits are generally placed side by side for ease of moving the cubicle and to save space, but this is not essential. The pit walls below the first 0.5 m are porous and the cover slab is tight-fitting.

Fly screen flush with top surface of the pipe and secured tightly

Air movement

Wind blowing over the top of the ventilation pipe causes air in the pipe to rise.

Replacement fresh air is drawn into the pit through the superstructure and down the toilet hole. This flow of fresh air keeps the superstructure free of odours.

Vent pipe

The VIP latrine differs from a simple pit latrine because it has a ventilation pipe and a more substantial superstructure.

Pit access handle

Flies

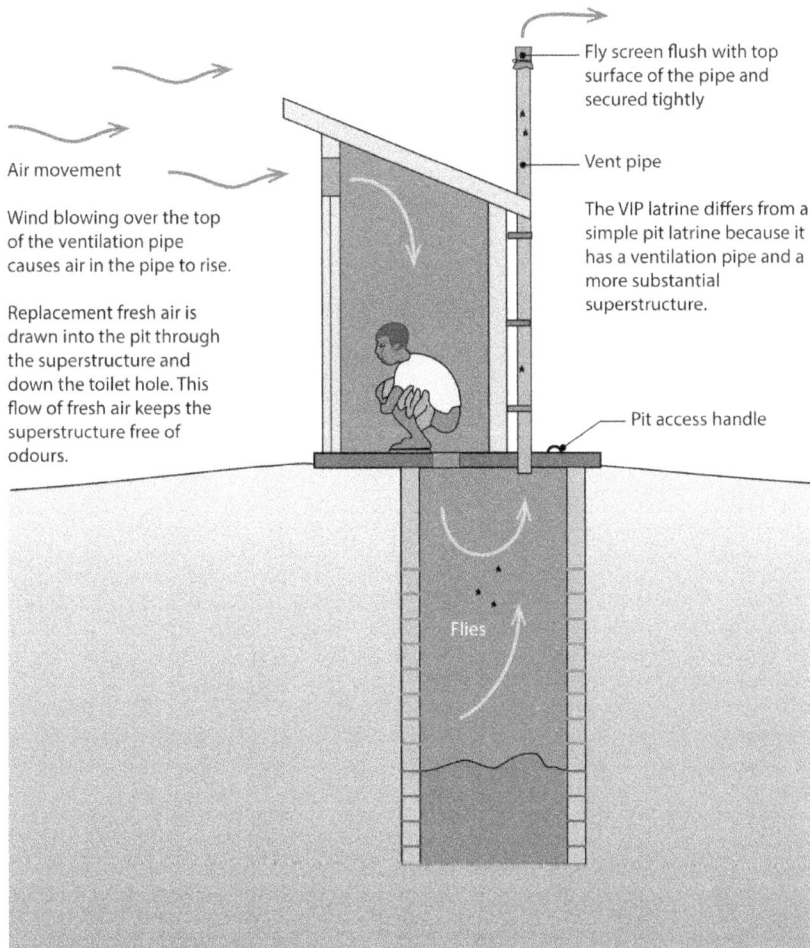

Figure 6.15 VIP toilet

The wall between the two pits is completely sealed to prevent cross-contamination. Parts of the cover slab of each pit should be removable to gain entry for emptying. See Chapter 6, section 2.4 for details of the design of the cubicle and ventilation pipe.

A defecation interface is placed over one of the defecation holes and a ventilation pipe installed over the corresponding vent hole. The holes in the other pit are closed and sealed. When the first pit is full, the defecation and ventilation holes are covered and sealed, the cubicle moved to cover the second pit, and the defecation interface and ventilation pipe installed. The second pit is then used until full, when the first pit is emptied ready for reuse. Provided the sludge has been stored in the pit for at least one year (but preferably two), it will be non-odorous and safe to handle. Depending on the materials used for anal cleansing, it may be

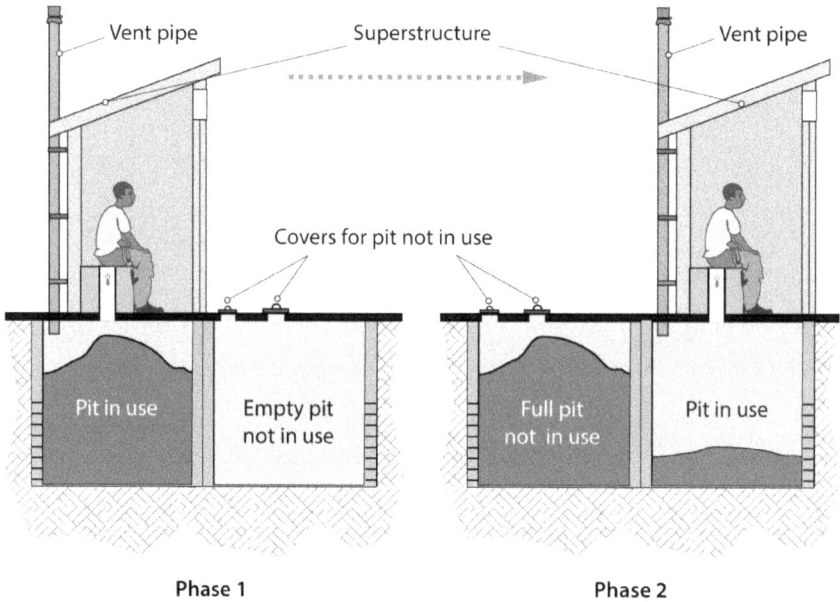

Figure 6.16 Direct entry twin pit toilet
Source: Adapted from (** Gensch, et al., 2018).

possible to dispose of the sludge locally; however, if non-biodegradable materials were used for anal cleansing, the waste may have to be considered as solid waste and disposed of appropriately (see Chapter 10, section 6).

Advantages
- Can be constructed above or below ground.
- May provide a safe method of disposing of human excreta locally if the toilet if operated correctly.

Disadvantages
- Expensive and complex to construct.
- Requires considerable user understanding of operation and maintenance.
- Requires an ongoing presence from toilet providers to support operation and maintenance.
- Has similar disadvantages to the standard VIP toilet.

6.2.6 *Vermifilter toilets*

Vermifilter toilets (VT), sometimes known as tiger worm toilets, (Oxfam Wash, n.d.b.) contain composting worms inside the pit that process and digest the faeces in situ, processing the raw sludge into vermicompost. This removes the need for traditional desludging, as the vermicompost is simpler to remove and builds up at a slower rate. This can lead to a reduction in the long-term operating costs and removes the need for expensive desludging and sludge treatment infrastructure. A worm colony can live inside the toilet indefinitely if the correct environmental conditions are maintained. There is very little experience of using vermifilter toilets in emergency settings.

Design and construction
The superstructure of a VT can be the same as existing traditional toilets, as long as there is a roof to prevent rainwater entering the system. As with all toilets, it is essential that the community is consulted regarding the design, location and sharing arrangements. To function correctly, the toilet must be constructed accurately and to a good standard. Poor construction will lead to operational difficulties and possibly a failure of the system.

In many places worms are grown commercially for use as fishing bait, for agricultural purposes to convert organic waste, or for the processing of sewage sludge through vermifiltration. Alternatively, worms may be found living in the local environment, particularly in areas with high amounts of organic matter, for example in areas where cattle gather. Sourcing worms in the quantities required for many toilets may be difficult or expensive, but once an initial batch of worms has been sourced it is relatively easy to grow more. Wormeries can be easily established to grow your own supply, but this adds cost and takes time – approximately six weeks to two months to double the stock of worms by weight.

Any worm suitable for composting is very likely suitable for a tiger worm toilet. There are several different species of composting worm that are widely available around the world but identifying the specific species of worm is difficult and is not necessary. Worms have various predators, including birds, snakes, small mammals, and invertebrates. Predators of relevance for VTs are mice, rats, and centipedes. Ants can aggravate the worms. Care should be taken to ensure that the toilet is well sealed to avoid predators entering the pit. Flies, maggots, cockroaches, and ants are not predators of composting worms.

A bedding layer is necessary for the worms to live in while they are in the toilet. A good bedding material should retain moisture, retain its porous (air-retaining) structure to prevent the toilet going anaerobic, and filter out the solids that are flushed. Possible bedding layers include woodchip, coconut fibre (coir), or compost. Whichever bedding layer is chosen, it should be soaked overnight before installation in the toilet and be added at the same time as the worms to a depth of around 100 mm. Key design factors are given in Table 6.4 and Figure 6.17.

Operation
Using the toilet is straightforward. After each use, add a cup of water to the pit.
Do not:

- use the toilet for bathing or laundry;
- throw non-biodegradable materials into the pit;
- use detergents or chemicals to clean the toilet as they will kill the worms.

Table 6.4 Key design parameters for tiger worm toilets

Pit surface area and weight of worms

Number of users	Weight of worms (kg)	Surface area of pit (m²)
5	1	0.7
10	2	>1.0
20	4	>2.0
30	6	>3.0

Liquid application rate

Minimum of 1.0–1.5 litres water (including urine) must enter pit with each use.

Environmental conditions for the worms

Parameter	Parameter Range	Optimum
Working temperature (°C)	5–35	20–25
Humidity (%)	50–96	60–80
Daily feed rate (kg feed/kg worms)	0.8–2.0	1.0
Food layer depth (mm)	–	100–150
Worm loading (kg/m²)	0.8–2.0	2.0

Source: Adapted from (Oxfam Wash, n.d.b.).

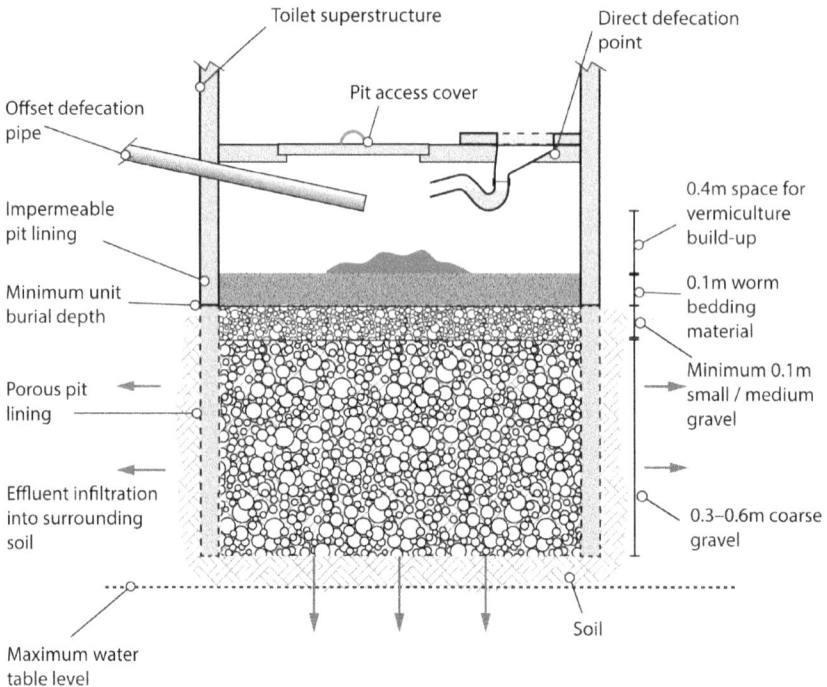

Figure 6.17 Tiger worm toilet
Source: Adapted from (Oxfam Wash, n.d.b.)

Advantages
- Emptying required only every 3–5 years.
- Costs a similar amount to alternative toilets but the reduced emptying period reduces operation costs.
- Less odour and flies than a pit toilet.

Disadvantages
- Unsuitable for water-scarce areas as flushing essential.
- Sourcing appropriate worms may be problematic in areas where composting is not practised.
- Unsuitable for areas with poor soil infiltration or where the water table is very high.
- Control of the volume of flush water used is critical, as overuse will kill the worms.
- Unsuitable for very cold or very dry climates.

6.2.7 Single-vault urine diversion toilets (UDTs)

Urine diversion toilets are commonly used for one of two reasons: in areas where subsurface infiltration is not possible and there is no infrastructure for a water-borne waste disposal system; or because of a wish to reclaim the nutrients in the urine/faeces for further use. Collecting the urine separately from the faeces can simplify their collection for off-site disposal or processing, reduces odour problems, and facilitates future nutrient reclamation. Also, in designs where the faeces and urine are collected in sealed containers, there is no contamination of the ground below. Typically, the user interface is divided into two compartments, one for faeces and the other for urine (Figure 5.13). Anal cleaning materials are collected separately. Extensive user education and mobilization is required prior to implementation. The main challenge is the sustainability of the collection and disposal infrastructure. The failure of this service has been the main reason for the widespread failure of this technology.

As the name suggests, both the urine and faeces collection tanks are located below a toilet cubicle (Figure 6.18). As the tanks become full, they are replaced and the full ones taken away for processing.

Design, construction, and operation
The defecation point either includes a urine diversion facility (see Chapter 5, section 3.3) or a separate urinal (male and female) and defecation point. Any wastewater from handwashing, bathing etc. must be disposed of separately.

The collection containers are housed in a chamber below the toilet cubicle, which, if possible, should be sited at ground level as this makes it much easier to access the containers for removal and replacement. The chamber may or may not be sealed depending on local circumstances. The sizes of the collection containers depend on the number of users and the frequency of

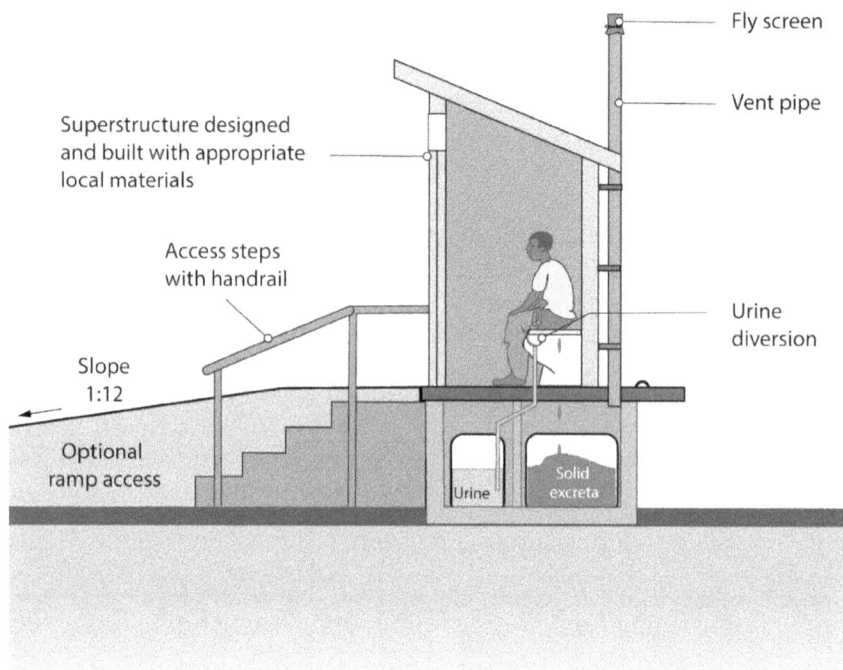

Figure 6.18 Single-vault urine diversion toilet
Source: Adapted from (** Gensch, et al., 2018)

emptying. Typically, the urine collection tank is around 5–20 litres and the faeces collection tank around 20–120 litres. In warm climates, small containers emptied frequently are preferred because of issues with fly breeding and strong odours from the urine and faeces. Also, small containers are easier to handle and transport. In very cold climates, containers tend to be larger as they may not be emptied for months. Whatever the size, containers must have tight-fitting lids to prevent spillage during transportation. Sometimes, where soil conditions allow, the urine is not collected but allowed to infiltrate into the soil below. This can lead to strong odours if a well-designed infiltration system is not provided.

The system requires the regular collection of used containers, which are replaced with empty, cleaned ones. The full containers are transported to the treatment site or transfer station where they are emptied and cleaned ready for reuse.

The theory is that the sale of the reclaimed nutrients (such as fertilizer) will pay most of the cost of the collection and treatment, with users paying a small contribution for the service. In practice, this is very rarely the case: insufficient funds are brought in by end-use sales and users are very unwilling to pay a high price for a service from which they get very little benefit.

Advantages
- The system does not pollute the ground below the toilet (unless urine is infiltrated).
- Valuable nutrients within the excreta are recovered for future use.
- Containers can be continuously emptied, offering a toilet of high value for money in the long term.
- Can be built of local materials.
- Suitable for areas with high groundwater, high-density building, areas subject to flooding, or rock close to the surface.
- High user acceptance subject to acceptable user fees.

Disadvantages
- Complex and expensive management service required for a sustainable system.
- Users must be educated in the use of the toilet, as wrongly mixing urine and faeces in the two tanks creates difficulties with collection and treatment.
- Does not dispose of wastewater from handwashing, bathing, etc.
- Very rarely sustainable in the long term, as the revenue does not cover the costs.
- In some cases, children may have trouble using the UDT properly and assistance is needed so their urine doesn't end up in the hole for faeces.
- Vault sizes are much smaller than in standard toilets and users may be put off by proximity to the faecal pile.
- Faecal sludge requires secondary composting to fully decompose and eliminate pathogens.

6.2.8 Double-vault urine diversion toilets

These are similar in concept to single-vault UDTs but without removable containers for faeces (Figure 6.19) (Section adapted from McBride, 2018).

Design, construction, and operation
These are similar to a twin pit direct entry toilet (see Chapter 6, section 2.5) but with a sealed floor and walls so that there is no infiltration of pollutants into the soil below. The urine must be collected in a container otherwise there is little difference between this and the twin pit direct entry toilet. If possible, the vaults should be constructed at ground level for ease of emptying and access. Good pit ventilation is essential to remove odours.

Vaults are used in turn, with the second one being used only after the first is full. Each vault should be large enough to hold the faeces of users for a minimum of a year but preferably two. When the second vault is full, the faeces in the first vault can be removed and safely disposed of locally as they are safe to handle and largely free of pathogens. The first vault can then be reused. When the second vault is full, the decomposed faeces from the first vault are removed and the vault reused. A vault size of 1.0 m³ (each vault) should be sufficient for a family toilet.

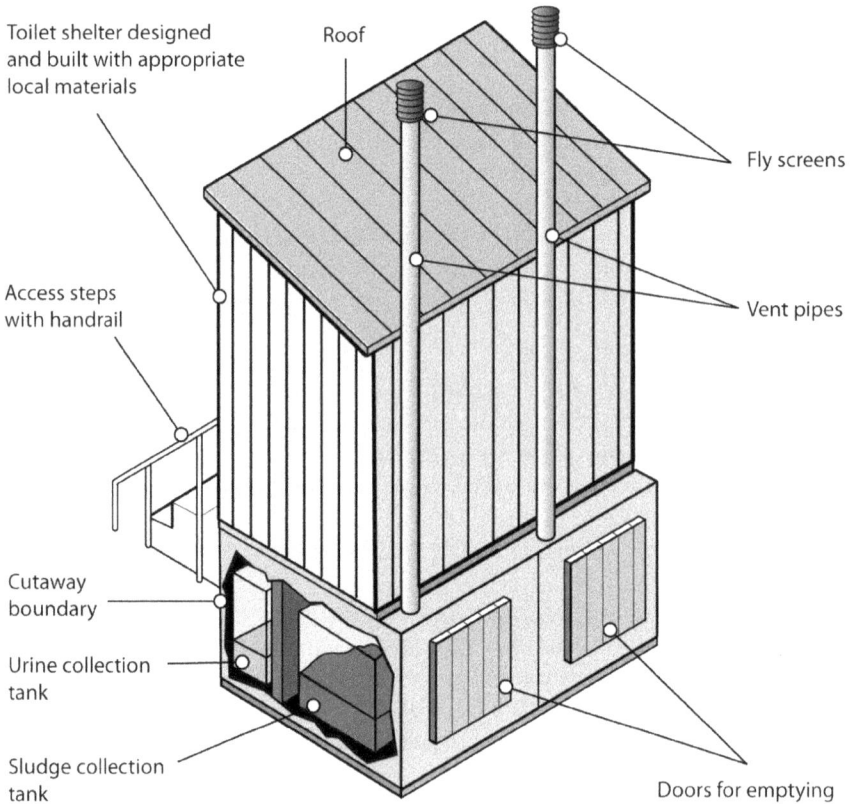

Figure 6.19 Double-vault urine diversion toilet
Source: Adapted from (** Gensch, et al., 2018)

   In general, the design has similar advantages and disadvantages to a single-vault UDT other than the following:

Advantages
   • Produces a dry decomposed and safe faecal sludge.
   • End-product useful as a soil conditioner.
   • No contamination of the groundwater.
   • Urine can be processed for nutrient recovery.

Disadvantages
   • Expensive and complex to construct.
   • Reliable management service required to collect and process the urine.
   • Unsuitable for high-density areas where there is no demand for soil conditioner.
   • Reluctance of users to handle or reuse excreta even if decomposed.

(Section adapted from McBride, 2018).

6.2.9 Vaults, cesspits, and holding tanks

Vaults, cesspits, and holding tanks are commonly used terms for a sealed tank that is used for storing fresh excreta.

Design, construction, and operation
The tank must be fully sealed and watertight. It must also have a tight-fitting cover and an independent ventilation shaft to remove odours. Their capacity is usually large enough for a few weeks' storage. They are used in high-density housing areas where water use is low, ground contamination is forbidden, there is limited space for an on-site treatment system or there is a lack of infrastructure (such as sewerage) for waste transport off-site.

Vaults (Figure 6.20) tend to be located directly below the toilet and store only excreta. Cesspits or holding tanks (Figure 6.21) are usually located adjacent to the building and collect all grey water via a domestic sewer system. They are often recommended when it is necessary to protect the local groundwater. The system allows for multiple water points within the building and a higher standard of sanitary facilities.

The critical component of any container system is the emptying mechanism. Vaults are frequently emptied manually into small tankers that may be drawn by hand or by animal. Cesspits, being much larger, are usually emptied mechanically. Anal cleaning materials must be limited to water or degradable toilet paper to prevent emptying problems. See Chapter 7 for more information on emptying techniques.

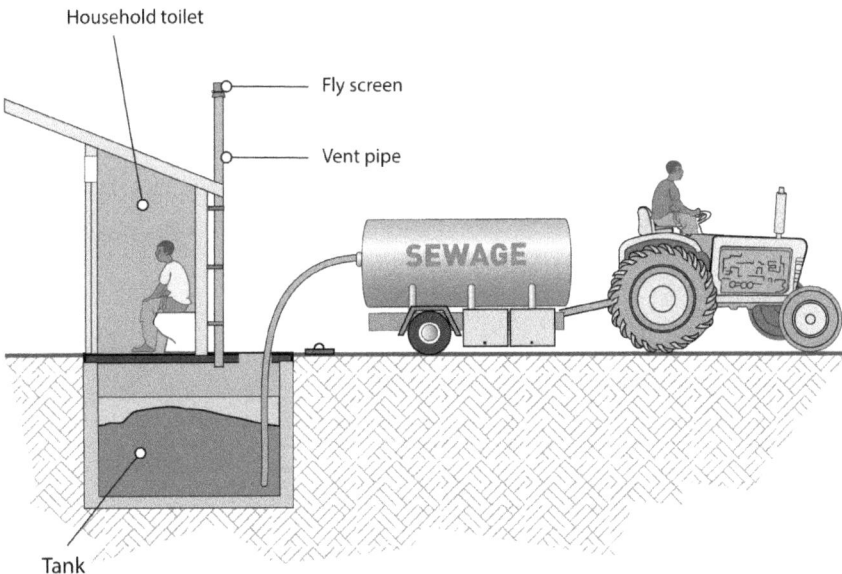

Figure 6.20 Vault toilet

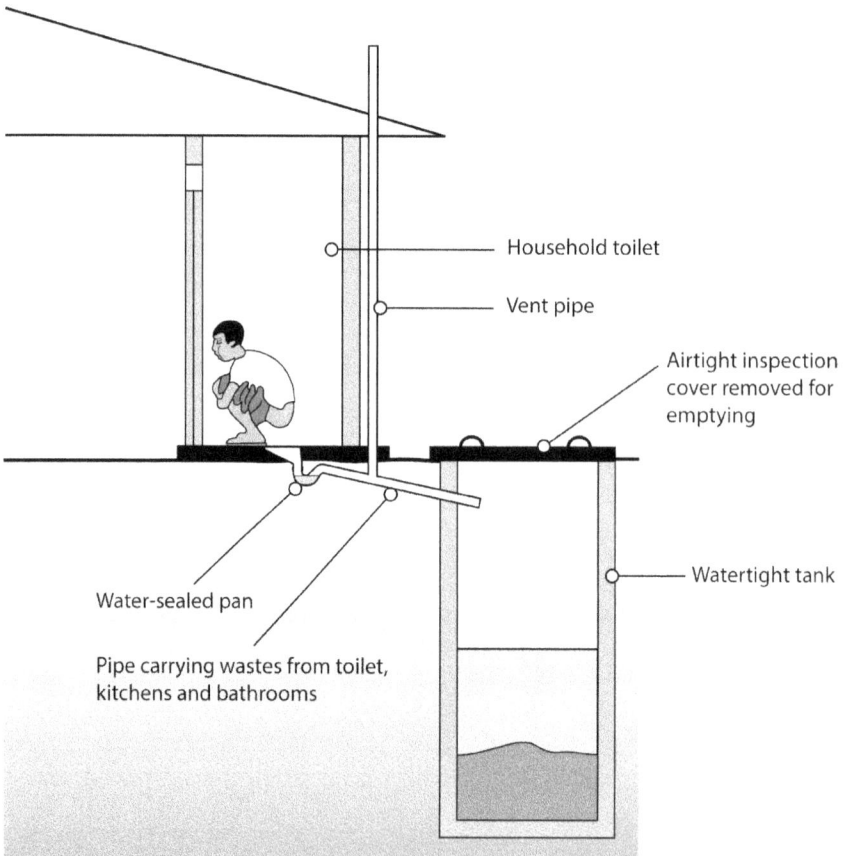

Figure 6.21 Cesspit/holding tank

Advantages
- The on-site components of the system are relatively simple and cheap.
- Suitable for very high-density areas where individual space is limited.

Disadvantages
- Complex and expensive off-site FSM services.
- Cess pits and holding tanks are suitable only for those who use water or soft toilet tissue for anal cleaning.
- Vaults do not dispose of wastewater or hard anal cleaning materials.

6.2.10 Dehydrating toilets

In hot, dry climates, dehydrating toilets dry the faeces, reducing their volume, minimizing fly breeding and odour, and reducing the pathogen load. In fully sealed units, there is also no contamination of the ground below. Dehydration

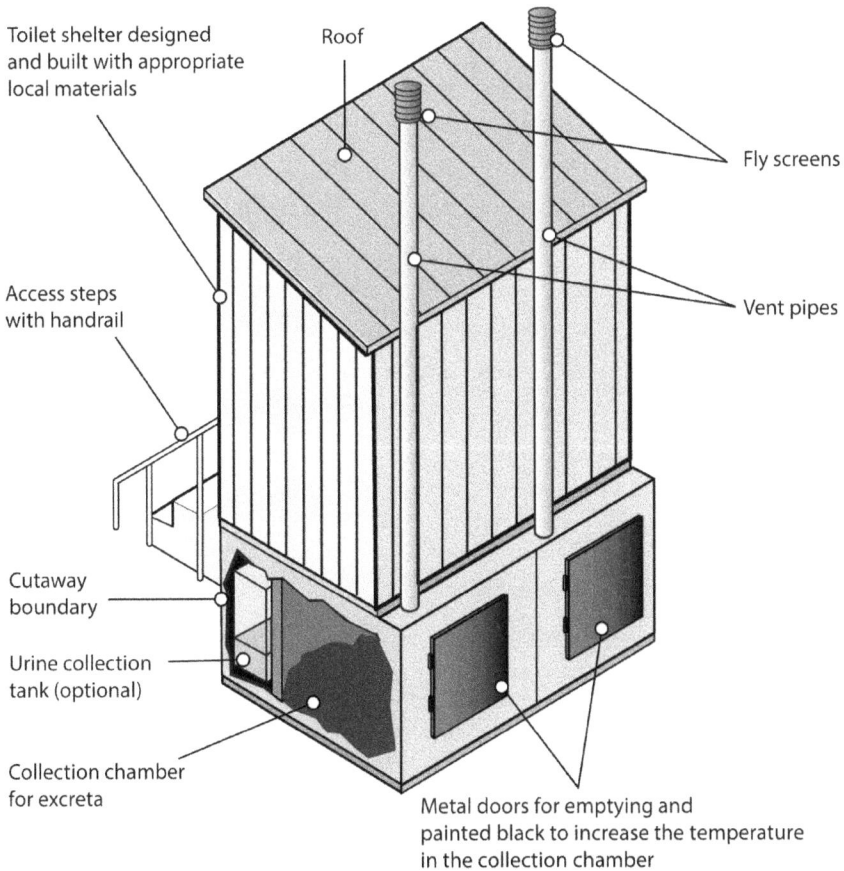

Figure 6.22 Double-vault dehydration toilet with urine diversion

toilets actively lower the water content of excreta without necessarily separating the urine from the faeces.

Design, construction, and operation
There are many different designs of dehydrating toilet, but the one shown in Figure 6.22 is the simplest. Two watertight vaults are constructed below the toilet floor, each having its own defecation point in the toilet. The part of the vault wall not inside the toilet cubicle is covered with a hinged, galvanized metal plate painted black. Each vault is designed to hold excreta and sometimes urine from all users for 1–2 years (about 1.0 m³ each vault). Each vault has a ventilation pipe to remove odours and water vapour. The whole unit is constructed so that the black metal plates face the sun.

The vaults are used one at a time, switching to the second vault when the first is full. The defecation point above the out-of-use vault is firmly covered. Used anal cleaning material should be collected separately and either buried or burned. The black metal plate on the vault absorbs heat from the sun that evaporates the urine and dehydrates the faeces. For that reason, a shallow, wide vault is better than a narrow, deep one. When the second pit is full, the dry faeces can be dug out and buried; they make a good soil conditioner.

Advantages
- Suitable for areas subject to flooding, a high water table, or in soils difficult to excavate.
- Does not contaminate the groundwater if urine is also collected.
- Suitable for those who use solid material for anal cleaning.

Disadvantages
- Unsuitable for those who use water for anal cleaning or where water use is high, unless collected separately.
- Toilets cannot be used for bathing, laundry, etc.
- Detailed user education essential.
- Unsuitable for humid climates.

6.2.11 Anaerobic composting toilets

Excreta is rich in nutrients that, under the right conditions, can be harnessed to produce an excellent resource that is typically used to produce fertilizer and flammable gas. In most situations, the production processes take place centrally, where large volumes of excreta make the conversion cost-effective. However, some processes can be undertaken at individual toilet level, even in emergencies.

Anaerobic composting toilets are very similar to double-vault UDTs (Figure 6.22), except that the vaults are larger. The difference is that the end product is a compost rather than a soil conditioner. More details on composting excreta are given in Chapter 10, section 6.4.

Design, construction, and operation
As with the double-vault UDT, two watertight chambers with ventilation pipes are positioned below a toilet cubicle that contains a defecation point into each chamber. Each chamber should be large enough to hold the combined faecal and organic wastes generated in a year. A chamber volume of round 1.5 m³ is usually sufficient for a single family and should take 1–2 years to fill. A layer of straw, sawdust, or similar is placed on the floor of one of the chambers, which is then used to collect faeces, organic waste (to control the chemical balance), and ash or sawdust (to control the moisture content). Urine is collected separately. When the vault is full, it is sealed and the second chamber used. When the second chamber is full, the contents of the first chamber can be removed ready for reuse. The process is called batch composting.

Batch composting toilets have had mixed success. Failure has generally been due to poor understanding of the process or lack of interest in the final product. Controlling the moisture content has often caused problems, particularly in communities where water is used for anal cleaning or people bathe in the toilet cubicle. Another issue is with not leaving the chamber mass long enough to properly decompose the waste and destroy the pathogens. There are examples of users, desperate for the fertilizer value of the compost, removing it after 3–6 months, which has led to outbreaks of faecal-related diseases.

Advantages
- Produces a dry decomposed and safe compost.
- Suitable for direct mixing with soil.
- No contamination of the groundwater.
- Urine can be processed for nutrient recovery.
- Suitable for areas with high groundwater, subject to flooding, or rock close to the surface.
- Chambers can be continuously emptied, offering a toilet of high value for money in the long term.
- Can be built of locally available materials.
- Suitable for those who use biodegradable materials for anal cleaning.

Disadvantages
- Expensive and complex to construct.
- Detailed education of correct operating procedures necessary to prevent system failure with consequent health risks.
- Reliable management service required to collect and process the urine.
- Unsuitable for areas where there is no demand for soil conditioner.
- Unsuitable for those who use water for anal cleaning.
- Those who use non-biodegradable materials for anal cleaning must dispose of them separately.
- Toilet cubicle cannot be used for bathing, laundry, etc.
- In dry climates, anal cleaning paper will not biodegrade, making the final mass unsuitable for use as a fertilizer without further treatment.

6.3 Toilets for institutions

Emergency toilets for institutions such as schools and healthcare facilities use the same technologies as for communal, family, and shared toilets, except that they are more robust to cope with constant use and there are more cubicles. Information on recommended cubicle numbers for schools and healthcare facilities are given in Table 5.2, but many factors must be borne in mind when designing any institutional toilet as shown in Box 6.4

There are numerous publications on the design of toilet facilities for schools and medical centres, such as (Mooijman, 2012) and (Noortgate, 2010) but very little on markets and bus stations. The principles are the same and can be adopted for a range of communal environments and users.

Box 6.4 Principle factors to be considered in institutional toilet design

- The toilet is public, therefore the users will feel no responsibility towards it. Accordingly, the facilities must be very strong, with a focus on simple design carefully chosen to meet the needs of a range of users and be easy to maintain. Commonly, some cubicles must be designed for the needs of specific user groups, as discussed in Chapter 5, section 1. Someone must take responsibility for supervising and funding the toilets' use, keeping them clean and maintaining the infrastructure. Unless otherwise formally agreed, that will be the organization funding their construction.
- Different user groups will have different expectations and needs from the toilet.
- Toilets in schools must accommodate the needs of teachers as well as those of children of various ages and requirements. It is common to provide dedicated toilet facilities for teachers. Issues of privacy for users and child protection are particularly important.
- Toilets in hospitals must accommodate staff and patients, many of whom may have a range of mobility and other issues.
- Bus station toilets typically have short bursts of high demand and require a higher than average number of cubicles.
- Markets also have periodic high demand during market days. They also tend to be robustly used and heavily soiled, so regular cleaning is essential.
- Privacy and security are common considerations. Careful door design, good lighting, appropriate screening, good circulation spaces in corridors and around handwashing areas, and good supervision are all essential to encourage people to use the toilets.
- Toilet access is important. Male and female toilets should be located separately, preferably not using common access paths. Small groups of toilets spread around the area being served is much better than a single large toilet block.
- Male urinals can considerably reduce the demand for toilet cubicles (see Chapter 5, section 3.5) and additional facilities in female toilets for menstrual hygiene management are essential.
- A reliable water source and a ready supply of soap are essential for promoting handwashing. In large toilet blocks this will generate high volumes of wastewater that must be disposed of appropriately.
- All public toilets that include a storage system such as a pit should be designed to be emptied from outside the cubicles. Generally, the volume of sludge to be removed is so large that mechanized emptying (Chapter 7) is required with associated vehicular access.

6.4   Maintenance and management of toilets

All toilets need regular cleaning and maintenance, emergency communal toilets especially so. Poorly maintained toilets are often the single biggest problem faced when promoting their use. Constant monitoring and supervision are required together with a sustainable supply of anal cleansing materials and cleaning products. Box 6.5 suggests the quantities of materials required for maintaining a communal toilet block. Large quantities of bleach or laundry detergent should not be poured down the interface if it discharges to a local septic tank as they may inhibit the natural biological degradation of the excreta.

Communal toilets should be cleaned at least once a day to prevent insanitary conditions, bad odours, and disease transmission. There may be cultural norms that affect the degree to which certain individuals in a society are prepared to be involved in the handling of faeces and the cleaning of toilets. This can affect the use and care of toilets and the recruitment

---

Box 6.5 Recommended monthly cleaning and sanitary supplies for a communal toilet block (six cubicles)

Liquid detergent: 20 litres
Bleach (5 –7%): 5 litres
Bucket: 1
Mop: 1
Stiff brush: 1
Handwashing soap: 4 kg (20 pieces)
Toilet paper: 200 rolls (if culturally appropriate)

Note: Quantities are for approximate guidance only and should be adapted to the context.

---

of sanitarians, toilet supervisors, and cleaners. Members of the affected community can usually be effectively employed through paid work or other incentives to undertake these tasks with proper supervision, equipment, and training. Be conscious of sustainability and the exit strategy when you choose to provide incentives for toilet-cleaning duties. If the organization taking over from you is unable to provide the same level of incentive, the toilet may quickly fall into disuse.

In some cases, it may be beneficial to engage community guards or patrols and deploy lighting at public toilets to ensure that the facilities are safe during the night. The organization of this service should ideally be defined and managed by the community itself, particularly women and girls, to ensure that the lighting and patrolling of the structures do not deter them from using the facilities. However, when determining the provision of lighting for communal toilets, ensure that there is an assessment of the lighting in the surrounding area. You should prevent only the toilet block being illuminated: often this can be a pull factor for people to hang out there socially in the evenings, thus limiting ease of use by women and girls.

Individual family toilets are nearly always the best solution as responsibility for maintenance is clearly understood. Failing that, the provision of toilets for small coherent groups is the next best option, with public toilets being the least desirable option. Whatever the choice, regular supervision is still necessary, especially where toilets are shared by a number of families, to ensure that the facilities are being looked after properly. Negotiations between users to smooth out interfamily issues are often required, especially where users were not previously known to each other. Hygiene promotion activities are crucially important to mobilize communities and to promote and ensure the cleanliness of toilets.

Lack of funds and the perceived temporary nature of an emergency situation make users unlikely to be able to fund or undertake activities that require major expenditure such as major repairs to the toilet, pit emptying, or downstream FSM activities. These roles must be undertaken by a competent organization or institution. Failure to assign and fund this role is another major cause of the failure of excreta disposal facilities.

Education should also be provided to the wider community to ensure that people are aware of the importance of using the toilets that have been provided and of corresponding hygiene practices, such as handwashing. Where there are toilets at health centres, particular attention should be paid to their maintenance and cleanliness as patients are likely to be more susceptible to disease. Toilets should not be used for the disposal of medical waste.

Here are some other key issues to consider when managing toilets:

- The disposal of non-faecal wasted into toilets is a major problem. Menstrual hygiene products, solid anal cleaning materials such as stones, magazine paper and corn cobs, organic and non-organic garbage etc. block pipes, fill pits and containers and make the operation of the remaining elements of FSM extremely challenging. Close supervision, education and the provision of facilities for the safe disposal of these items are all necessary.
- Clarify and agree with partners who owns what. By default, the agency that built the infrastructure owns it and is therefore responsible for operation and maintenance (O&M). However, families are generally expected to take responsibility for day-to-day operation of their own family toilet. Early consultations with community leaders, users, and private contractors may be able to arrange O&M services for the longer term but there is likely to be a financial implication. In all cases, an organization is required to provide management and support to any excreta disposal service. This must be agreed before construction work begins.
- Coordination with other agencies working in the same area is important to ensure that a consistent approach is adopted. If people in one location are paid for O&M and people in another location are expected to perform the same tasks on a purely voluntary basis, this is likely to create unrest.
- For large sites, such as major camps, the sheer volume of work required for appropriate O&M is huge. This makes the scale of supervision difficult. It is important that community members are empowered to manage this wherever possible (normally at a cost).
- The quantity of equipment required for cleaning (household bleach, mops, rags, etc.) may also be considerable (see Box 6.5) and an appropriate distribution system must be developed. This is commonly implemented in conjunction with a hygiene promotion programme.

CHAPTER 7

Toilet emptying

Emptying and transporting faecal sludge should be undertaken only as a last resort (such as in urban or high-density settings, hard ground, or high water table contexts) where no other on-site faecal management option is possible. Desludging and transportation operations can be very complex and costly, and there should be a clear exit strategy from the start. Moving faecal sludge around can pose a major risk to public health and the environment if it is not undertaken properly.

There is a wide range of technologies for emptying faecal waste but virtually none of them handle gross solids such as cloths, plastic bottles, slow degradable paper, sticks, sanitary napkins, and other solid waste. Eliminating and reducing the quantity of these items entering the toilet has a major impact on the efficiency and effectiveness of emptying technologies.

Chemicals or oil are often added to faecal waste to control odour, reduce fly and mosquito breeding, and increase sludge decomposition rates. These are not recommended where sludge is expected to be transported for future treatment as they could affect downstream treatment processes.

An overview of different emptying methods is given in Table 7.1 which summarizes the advantages and disadvantages of each.

7.1 Manual emptying

The most basic method for emptying tanks and pits is with a bucket. If the wastes are primarily liquid, emptying can be achieved from outside, lowering and raising the bucket on a rope. However, this is not possible if the wastes are solid or semi-solid or contain solid materials such as cloths or garbage. In some countries, it is common for workers to enter the pit to dig out such wastes. **But under no circumstances should a person be expected or permitted to enter a pit containing fresh or semi-decomposed faecal sludge for the purpose of emptying unless provided with full personal protective equipment (PPE), including breathing equipment and a managed safety harness. The risks to the worker's life, health, and well-being are too great for any emergency faecal sludge management organization to accept.** Manual emptying should always be seen as a method of last resort and a short-term measure to be replaced as soon as possible. There are no specific design parameters for manual emptying. The bucket should preferably be made of metal rather than plastic because the weight makes it easier to fill. There should always be at least two people at the top of the pit in case there is an accident and someone falls in. If possible, the person emptying the pit should wear a harness attached to a secure point in case they fall. Workers should also be trained in what to do in the event of such an occurrence. Full buckets should be emptied into a container situated adjacent to the pit. Carrying full buckets should be avoided because of environmental contamination from spillage. When emptying is complete, the area around the pit should be sluiced down with chlorinated water and the wastewater directed into the pit.

End-of-shift bathing facilities, clean clothing, and cleaning of contaminated clothes are essential for all workers involved in this activity.

If the wastes are not liquid, or contain a large quantity of non-faecal material, manual emptying is more difficult. Entering the pit is highly dangerous because of the unstable nature of the material and the high possibility of poisonous gas being present. It may be possible to liquefy the wastes by adding extra water and stirring the contents but if that is not possible the best answer is to abandon the pit and either fill it in with soil or seal the top and leave for at least a year (preferably two) before digging out the contents.

7.2   Hand-operated machines

Replacing the bucket with a hand-operated pump eases the emptying process and reduces contamination of the workers and the local environment. A range of hand-operated machines for emptying pits have been developed around the world with mixed results.

Most designs are based on the simple lift pump, but there are some that use a positive displacement mechanism and others that rely on the helical screw. More detailed information on the options can be found in O`Riorden, 2009. Figure 7.1 shows three of the hand-operated treatment technologies

Figure 7.1 Hand-operated emptying machines
Source: Adapted from (EAWAG, Sandec, 2008) & (Georges, et al., n.d.)

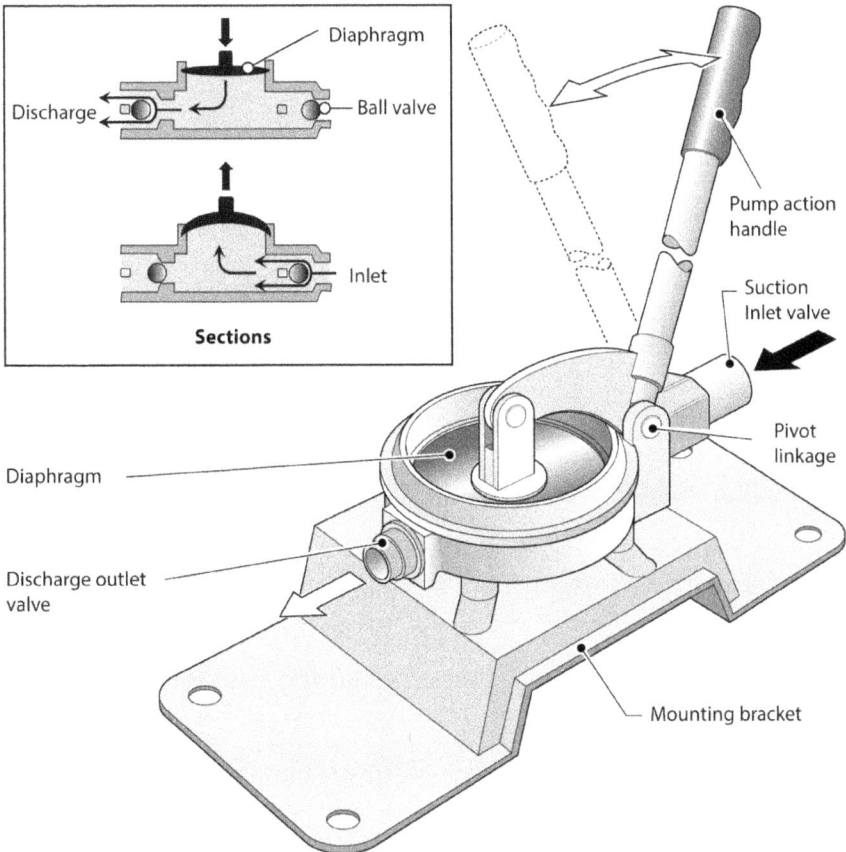

Figure 7.2 Principles of a diaphragm pump

available. Most will handle only liquids or semi-liquids and none can handle gross non-faecal solids. In some cases, dry faecal sludge can be lifted by adding water and agitating the contents to make it semi-liquid, but that increases the volume and weight of sludge to be transported and treated.

Hand-operated diaphragm pumps can be adapted for use with a power source for increased capacity (Figure 7.2), they are a relatively lightweight, and rugged technology that can pump dense slurry. They are simple to operate and can run dry, but are high maintenance and inefficient.

7.3 Centrifugal pumps

Centrifugal pumping units are similar to portable water pumping units but must be fitted with an open impeller capable of handling small solids. There are many different designs of centrifugal pump impeller but Figure 7.3 shows the basic difference between open and closed impellers. It is preferable that they are self-priming as this makes use easier and prevents the pump

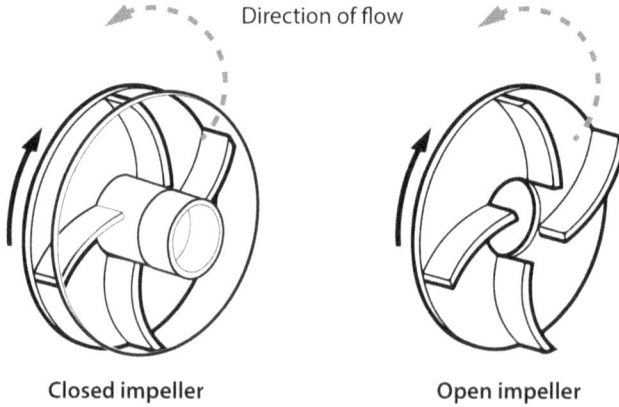

Direction of flow

Closed impeller   Open impeller

Figure 7.3 Open and closed centrifugal pump impellers

running dry (Figure 7.4). They can empty faecal contents much more rapidly than manual methods and can transport the slurry a greater distance. Total maximum physical lift (difference in elevation between the surface of the slurry in the pit and the inlet to the pump) should be limited to 2.0–3.0 m provided the length of the suction pipe is less than 30 m (** Gensch, et al., 2018). Pump manufacturers publish the maximum size of solids their pump can handle but, in the absence of such information,

Figure 7.4 Pit emptying using centrifugal pump

the size should be limited to approximately one-quarter the diameter of the pump inlet. The suction pipe inlet should always be covered by a coarse screen to prevent the entry of gross solids that would block the pipe or damage the pump.

Centrifugal pumps are normally powered by a close-coupled petrol or diesel engine for ease of portability, but in areas with good vehicular access the pump may be mounted on a specialized emptying vehicle with storage tank.

7.4   Modified powered auger

Auger pumps use a helical posthole drill fitted inside a uPVC pipe to lift the sludge (Figure 7.5). They are one of the few technologies that can lift dry sludge as the drill bit is in direct contact with the sludge. Auger pumps must be powered but the rotation speed should be restricted to 60–90 rpm (GOAL, 2016). The outer casing must be easy to remove in case of blockages. Although the auger can cope with more garbage than other pumps, is simple to use, and reaches relatively high flow rates, it is also very heavy (up to 40 kg when emptying), is difficult to clean, and has a fixed length. Collecting the sludge as it discharges from the pump can also be difficult as the outlet is constantly moving up, down, and around the pit.

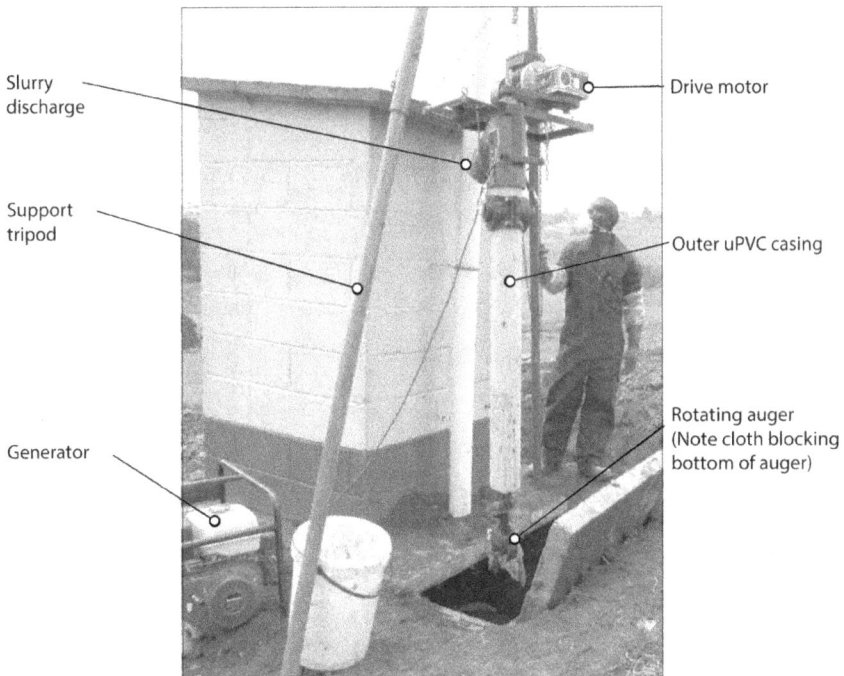

Figure 7.5 Modified powered auger
Source: Adapted from (Rogers, et al., 2014)

7.5   Vacuum suction

Vacuum systems have the big advantage that there are no valves or moving parts between the inlet pipe and the storage tank. This means that they can pump a wide range of semi-solids without becoming blocked. They are poor at pumping dry material and so it is common to add water and agitate the sludge prior to pumping. The operating principle is very simple: an air pump lowers the pressure in a sealed tank. A pipe connected to the tank is placed in the stored faecal waste and the pressure difference between the atmosphere and the sealed tank forces the slurry into the tank (Figure 7.6). Nearly always, the vacuum pump and storage tank come as an integrated unit, normally mounted on a vehicle. Most commercial vacuum sludge trucks have storage tanks of between 10,000 and 15,000 litres for heavy-duty urban environments. It is rare to find small units suitable for emergency response, such as the example shown in Figure 7.7.

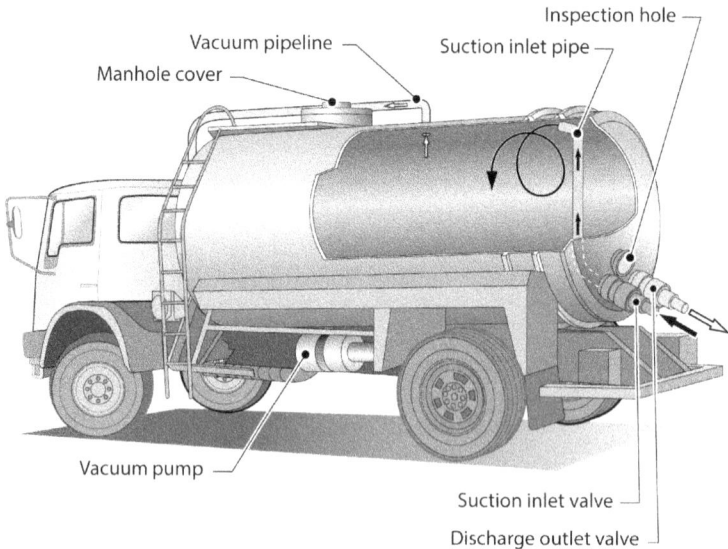

Inspection hole
Vacuum pipeline
Suction inlet pipe
Manhole cover
Vacuum pump
Suction inlet valve
Discharge outlet valve

Figure 7.6 Operation of a vacuum tanker

Typical commercial vacuum tanker

Small vacuum tanker for high-density housing areas

Figure 7.7 Vacuum tankers

Table 7.1 Advantages and disadvantages of emptying options

Lifting device	Approximate removal rate	Advantages	Disadvantages
Manual	—	• The only feasible method for removing gross solids. • Simple to establish. • Uses locally available tools and unskilled workers. • Does not require vehicular access. • No external power required. • Minimal maintenance costs.	• Unhygienic with frequent spillage of faecal waste around the top of the pit. • Exposes workers to faecal-related infections unless full PPE is provided, good working practices enforced, and bathing and laundry facilities provided. • Often liberates strong odours.
Hand-operated machines	—	• Many designs can be fabricated locally using basic skills and materials. • Suitable for sites with limited access. • When properly managed, leads to less spillage than manual emptying. • Lower health risks for operators than manual emptying.	• Hand-pumping faecal sludge is hard work. • Lifting mechanism frequently blocks, especially when non-faecal solids such as cloths, garbage, or solid anal cleaning materials are present. • Equipment is not robust and prone to frequent breakdowns (often the diaphragm).
Centrifugal pumps	750–1,500 litres/m	• Suited to faecal material with high water content (e.g. the contents of chemical toilets or septic tanks). • Depending on the pump characteristics and the power source, large volumes of waste can be pumped to high elevations and over significant distances.	• Unless the sludge is constantly agitated, they tend to remove the liquid portion of the waste first, leaving the thicker sludge behind. • Easily blocked by plastic bags, stones, leaves, sanitary pads, etc. • Pumps can easily be damaged by running dry unless designed to be self-priming.

(Continued)

Table 7.1 Continued

Lifting device	Approximate removal rate	Advantages	Disadvantages
Centrifugal pumps (cont.)	750–1,500 litres/m	• Equipment easily sourced at low cost. • Units highly portable. • Some models are self-priming.	• Inefficient and high maintenance. • Low discharge rates and limited total pumping head (approximately 7 m).
Diaphragm pumps	100–300 litres/m	• Highly portable, simple, and robust. • Capable of pumping thick slurry. • Can be allowed to run dry. • Can accommodate larger solids (but should still be protected from gross solids). • Self-priming if properly serviced.	
Modified power augers	25 litres/m @ >60 rpm	• Pumps dry sludge including some gross solids. • Locally fabricated from readily available parts.	• Heavy to use. • Typically only lifts sludge up to 3 m. • Unconfined discharge leads to local contamination. • Low discharge rates. • Does not lift liquid or low viscosity slurry. • Frequent blockages.
Vacuum suction	500–1,500 litres/m	Suited to faecal material with high water content (e.g. the contents of chemical toilets or septic tanks). Very little environmental contamination around the collection point.	The intake, filter, pump, and exhaust units are examples of high-precision engineering and require frequent attention to keep operating. Most units are large and unsuitable for accessing narrow roads and alleyways.

(Continued)

Table 7.1 Continued

Lifting device	Approximate removal rate	Advantages	Disadvantages
Vacuum suction (cont.)	500–1,500 litres/m	Includes transport of faecal waste to treatment site.	Manufacture is slow (6 months plus), making the reliance on new units in an emergency impractical.
		Frequently commercial companies are readily available to take on the task at short notice.	Very expensive and generally unsustainable.
			Spare parts difficult to source.
			Easily blocked by plastic bags, stones, leaves, sanitary pads, etc.
			Suction lift limited to 6 m at best, often with no positive pumping capacity above the level of the vehicle.

CHAPTER 8

Faecal sludge transport

Any toilet that requires desludging will also require some form of faecal sludge transport system. This is typically performed using mechanized or animal-drawn vehicles or manual cartage. However, in some urban areas it may use underground pipes (sewers). Regular desludging and transport may be the only possible option in urban areas or sites with hard ground, with shallow groundwater, or subject to flooding. The distance transported varies greatly from a few metres to many kilometres. In any case, it can be a risky and costly business. Faecal sludge managers must ensure that all operations are carried out by professional and qualified organizations with adequate safety and public health monitoring protocols. Care must be taken to ensure that operations do not pose health risks to the population or the personnel delivering the service, nor environmental contamination.

8.1 Manual carriage

This is the oldest method of transporting faecal waste and is still common in emergencies, especially over short distances. Despite that, it should be avoided wherever possible because of the high risk of environmental contamination and workers' exposure to communicable diseases. Containers used for hand carriage must have a tightly fitting lid and be ergonomically designed to suit the method of carriage. The operatives must be provided with appropriate personal protective equipment (PPE). Larger containers can be carried by multiple operatives (Figure 8.1). This method reduces spillage and enables carriage over longer distances.

The amount that a person can carry varies from person to person and according to the shape of the container, the distance carried, the terrain over which it is carried, and the position on the body that carries the weight. General guidance is given in Table 8.1. In emergencies, manual carriage is usually seen as a temporary response until a longer-term solution can be implemented; in urban settings, it may already be the established means of transporting faecal waste.

Advantages
- Quick and simple to establish, low cost, and uses unskilled workers and local materials.
- Highly suited to unplanned settlements with narrow pathways.
- Provides an income for members of the affected community.
- Often an established transport method for faecal sludge in urban settings.

Figure 8.1 Four men carrying a 100-litre container of faecal sludge

Table 8.1 Approximate weights and distances for manual carriage of faecal sludge^

Carriage method	Maximum weight (kg)	Maximum carriage distance (km)
Single person by hand	20	0.5^^
Single person on shoulders or back	25–35	0.5
2 people	40–50	1–2
4 people	80–100	1–2

Note: ^Assumes healthy individuals of average size wearing appropriate clothing and using containers and carrying devices specifically designed for the task.

Source: ^^Based on SPHERE indicators for access to water points (** Sphere Association, 2018).

Disadvantages
- May be considered a degrading activity among the user population.
- Very difficult to manage, particularly with a large number of teams.
- Unsuitable for low-density settings.

8.2   Manual and animal-powered vehicles

A more efficient form of carriage is using a small cart powered either by a human or an animal. Such vehicles are common in less developed countries and most effective where there are graded and level access roads. Table 8.2 suggests maximum carrying capacities for manual and animal-powered carts.

The faecal sludge is normally stored in multiple small containers, such as buckets with lids, or a watertight tank. These must not pose a public health or environmental risk to the community. Many poorly managed

Table 8.2 Maximum load limits for manual and animal-drawn carts

Device	Max load (kg)	Comments
Manual, 2-wheeled hand cart	114	Maximum distance per day 3.2 km^
Manual, 3-wheeled hand cart	227	Maximum distance per day 3.2 km^
Manual, 4-wheeled hand cart	227	Maximum distance per day 6.6 km^
Donkey-pulled cart	450	Twice body weight of donkey

Note: These figures assume healthy humans and animals of average size pulling a well-designed and constructed vehicle over a smooth flat and level surface.

Source: ^(Canadian Centre for Occupational Health and Safety, n.d.)

transportation solutions in less developed countries use simple open-topped carts. While not recommended, these may be a temporary measure provided the sludge is dry. Wet sludge will leak strong liquor onto the ground as the cart passes through the community. Where the sludge is liquid, the vehicle must be designed to withstand the lateral forces exerted by the liquid as it moves from side to side. Containers must have tight-fitting lids and large containers may need internal baffles to reduce liquid movement during transport. Vehicles used for faecal sludge transport should be examined and approved prior to being used.

Advantages
- A relatively low-cost solution using local resources and accepted practices.
- Effective over relatively short distances.

Disadvantages
- Difficult to manage.
- Vehicles frequently poorly maintained and spill faecal sludge.
- Cultural differences in attitudes to animal welfare can be difficult to overcome.

8.3   Mechanized vehicles

Where faecal sludge is transported by dedicated mechanized emptying vehicles such a vacuum tankers (see Chapter 7, section 5) it is usual for the same vehicles to be used for transport however, standalone mechanized emptying devices such as those described in Chapter 7, section 3 require separate facilities for transporting the sludge. Vehicles can be specifically designed for the transport of faecal sludge (Canadian Centre for Occupational Health and Safety, n.d.) or a basic flatbed truck can be fitted with a rigid or flexible tank (Figure 8.2). It is common to fit any vehicle carrying faecal sludge with a pump to speed up filling, emptying, and recirculation to prevent sludge settlement, as shown in Figure 8.3.

Figure 8.2 Trailer fitted with flexible storage tank for transporting faecal sludge
Note: Trailer side panels normally raised during transport when the flexible tank is full.

Figure 8.3 Mechanized faecal sludge transport using a tractor with slurry trailer and pump

Advantages
- Rapid, hygienic, and flexible method of transporting large quantities faecal sludge over long distances.
- Uses readily available private transport companies.

Disadvantages
- Expensive for long-term use.
- Difficult to manage and open to corrupt practices.
- Flexible tanks require careful restraint: large volumes of unrestrained liquid are dangerous and can damage the vehicle or affect steerage.
- Flexible tanks are very difficult to clean.

8.4   Sewers

8.4.1 Existing sewer networks

Many urban areas have existing sewer networks that, where possible, should be used for disposing of faecal sludge. Many existing properties will already

---

Box 8.1 Checklist for assessing an existing sewer network

The primary objective of using a sewer network in an emergency is to safely transport faecal waste to somewhere it cannot infect people. Transport to a functioning treatment plant is an added bonus but not the priority. The assessment process depends on whether a functioning sewerage management authority (the utility) exists or not.

A functioning sewerage utility exists
- Does the utility have the capacity to assess the functionality of the network post-emergency and carry out essential repairs? If so, the utility should take the lead.
- Would the provision of additional equipment, financial, or technical resources help them carry out necessary repairs? This is commonly the case. Utilities are often chronically under-resourced and additional external support will be necessary to repair the network.
- Does your organization have formal permission to make alterations or additions to the network as part of the emergency response? Public utilities generally have a legal responsibility for the management of sewer networks and will insist on authorizing agencies that want to work on them.

A functioning sewerage utility does not exist
- Are there any previous sewerage employees still living locally who are willing to assist in the operation and repair of the network? Previous employees often have extensive knowledge of the network and can be a major asset in assessing the current status and operation of the system.
- Is there a functioning water supply that is sufficient to keep the sewer network operating? Sewers require a constant water flow to transport faecal wastes along the network. For typical sewer diameters, a minimum peak flow velocity of 0.6–0.7 m/s is recommended. This may require additional water to be added to the sewer. Water tankers can be used to flush sewers once or twice a day to prevent a build-up of solids in the network.
- Is the network damaged in any way – for example, breaks, leaks, or blockages that will hinder the sewage flow to a safe discharge point? Start checking the system from the sewer outfall and work upstream. If a major breach is found, consider, as a temporary measure, breaking the network into a number of smaller systems, each discharging to a surface draining network or some other body such as a lagoon or temporary treatment plant. An assessment of the route of the drainage channel, its possible health implications, and the removal of any obstruction to the flow must be done prior to making a decision

---

have the infrastructure and connections to the system and it is often a simple task to connect new temporary buildings constructed close by. Prior to use, the sewer network must be assessed to check its functionality and capacity to accept additional volumes of faecal waste. Box 8.1 provides a checklist for assessing the condition and suitability of an existing sewerage network. Once a section of the sewer network is functioning satisfactorily, existing toilets can be rehabilitated and reconnected. Temporary toilet blocks can be added to meet additional needs, such as the one shown in Figure 6.7 or the block shown in Figure 6.6 (connected to a sewer rather than a storage tank).

8.4.2 Sewage pumping stations

Many sewer networks include pumping stations to lift the sewage to a higher level either to overcome local geography or as part of the treatment plant system. Work closely with the local utility or, if that is no longer functioning, its local employees, as they have an in-depth knowledge of how the pumping stations work (and often how to fix them). Most pumping stations are designed with an emergency overflow in case of mechanical or electrical breakdown. This is commonly a connection to an existing surface water drainage network. In the first instance, check for and clear any obstructions in the emergency overflow system so that the sewage can drain away without flooding homes or important infrastructure.

Assess the reasons for the pumping station failure. It may be something that is within the competencies and available equipment of the local utility or assessing organization. However, especially in large pumping stations, a full assessment by specialist agencies may be necessary and a decision taken as to whether that is appropriate at the current stage of the emergency response. In the short term, the primary objective will be to keep the emergency overflow network functioning effectively and to minimize potential health risks.

8.4.3 Emergency sewer systems

In some cases, temporary sewers may be required to connect toilet facilities to a storage tank or an existing sewer network. Occasionally, it will be necessary to install a complete sewer network, such as when establishing a new refugee camp. This section only contains advice on the design and construction of short, temporary sewers. The design of sewer networks is beyond the scope of this book and professional advice should be sought. Many countries have codes of practice for the design and construction of sewer networks. These codes cover such items as minimum pipe diameters, maximum and minimum gradients, sewer pipe materials, and minimum invert depths. However, emergency sewer networks are not the same as standard installations:

- They are installed rapidly with minimum time for detailed design and environmental considerations.
- They are usually considered temporary infrastructure (although it is common for them to be in use for much longer than anticipated).
- Flow patterns, volumes, and sewage components are different from normal applications.

Locally approved codes of practice should be used as much as possible, as these will be better understood by local technical staff and be more likely to be approved by authorities. The guidance given in this section is based on UK codes of practice (Ministry of Housing, Communities & Local Government (UK), 2010); however, this will tend to oversize sewers because of higher water use in the UK than in most refugee camps. More information on emergency sewer design with some worked examples is given in Appendix 4.

8.5   Transfer stations

In many cases, it is not possible to transport faecal sludge directly to the treatment or disposal facilities; an intermediate transfer station is often required. Transfer stations fall into two broad types: sludge transfer from

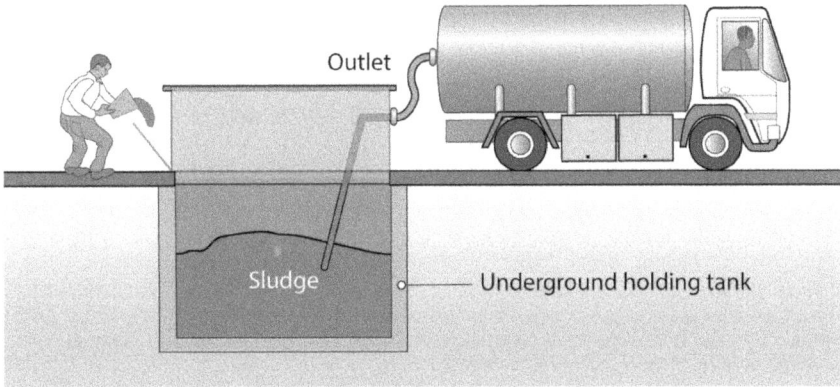

Figure 8.4 Small to large-scale transfer station
Source: Adapted from (** Gensch, et al., 2018)

numerous small collection vehicles (such as hand carts) to a larger vehicle; and sludge discharge from a vehicle to a sewer network.

Vehicle-to-vehicle transfer stations are commonly provided in high-density housing areas where large vehicle access is impossible and the treatment facilities are a distance from the serviced area (Figure 8.4). The detailed design will depend on the topography and geography of the site and the design of the incoming and outgoing vehicles. In general, an intermediate storage tank will be required to balance the inlet and outlet flow rates; this can be any suitable container that meets the local needs. Some key issues to consider when designing the station are given in Box 8.2.

A temporary mobile transfer station can be set up while toilet pits in the area are being emptied. Such mobile transfer stations consist of easily transportable containers or vacuum tankers, temporarily located at a site where multiple trips by small-scale transport equipment is necessary because of the density of the settlement. The faecal sludge can be sucked from the smaller vacuum tankers, large drums, or buckets using a portable slurry pump and discharged into a larger vehicle for onward transport. When all the pits in an area have been emptied, the transfer station can then be relocated to a new area.

Box 8.2 Key design considerations for designing a transfer station

- For manually hauled faecal sludge, use Table 8.1 and Table 8.2 for planning the positions of transfer stations.
- The site should be away from housing areas because of the strong odours produced by the transfer process.
- The size of the holding tank is a function of the rate at which sludge is delivered to the transfer station and the size and frequency of the bulk transport vehicle (see example below).
- The storage tank must be watertight and vented to prevent environmental contamination and gas accumulation.
- Facilities for stirring the sludge may be necessary to prevent settlement of heavy solids.

(Continued)

Box 8.2 Continued

- A large amount of sludge spillage is common. Working surfaces must be fully sealed and graded to drain to a safe collection and treatment site.
- A reliable water supply is required for sluicing the working area.
- The whole site should be securely fenced.
- Health and safety facilities, including full PPE, and bathing facilities for workers are essential

Example
Four teams of manual faecal sludge haulers deliver 2 m³ of faecal sludge to the transfer station every hour during an eight-hour shift. The sludge is transferred to a treatment plant in a 6 m³ tanker every four hours. What size of holding tank is required if the tanker can work a 12-hour shift? Assume the holding tank is empty at the beginning of the day.

Hour from starting work	Incoming FS volume (m³)	Volume in storage³	Tanker collection volume³
1	2	2	
2	2	4	
3	2	6	
4	2	2	6
5	2	4	
6	2	6	
7	2	8	
8	2	4	6
9	0	4	
10	0	4	
11	0	4	4
12	0	0	

The minimum size of the holding tank is 8 m3, although add 50–100 per cent extra for incoming volume variation, vehicle late arrival, and prevention of splashing during filling and emptying.

When transferring sludge to an existing sewer, do not use an existing access chamber: the opening is too small and there will be a large amount of spillage. It is more appropriate to construct a dedicated discharge point following the guidelines given in Box 8.2. A general layout of a sewer transfer station is shown in Figure 8.5. The coarse screen is to prevent gross solids entering the sewer systems where they may cause a blockage. The screen is particularly important where sludge has been manually emptied and transported. The area around the transfer point and where vehicles stand must be fully sealed and graded towards the inlet. An adequate water supply is required for sluicing down vehicles and the transfer area. The station must be fenced for safety and security.

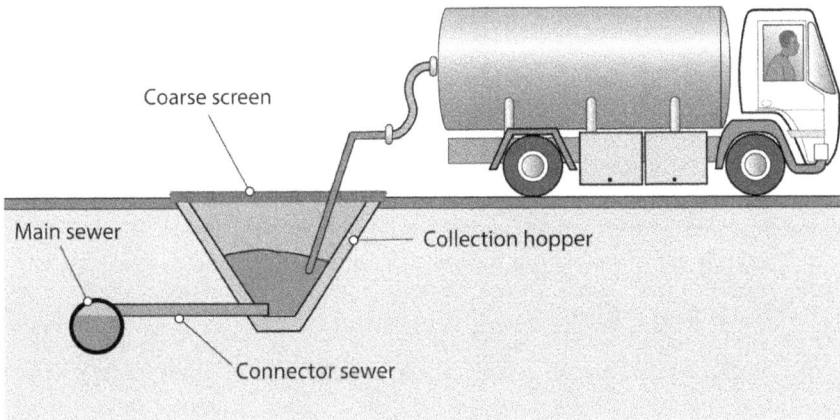

Figure 8.5 Transfer of faecal sludge to a sewer

Faecal sludge typically contains a high proportion of grit that can cause added problems in the sewer network, pumping stations, and treatment plants. Sludge should be added to large-diameter trunk sewers only where the flow volume is high enough to keep the grit in suspension. Sludge that has been collected manually will commonly require the addition of water to lower the viscosity and ease the flow into the sewer network. Transferring sludge to a sewer that discharges to a pumping station should be avoided as the grit will settle out in the pumping well and possibly damage the pumps.

Faecal sludge has different biochemical characteristics to typical sewage. The sudden arrival of a large amount of faecal sludge at a treatment plant may have significant effects on the biological processes being used. Approval from treatment plant managers must always be obtained prior to establishing a new faecal sludge discharge point into a sewer network.

Advantages
- Reduces transport distance and makes sludge transport to the treatment plant more efficient, especially where small-scale service providers with slow vehicles are involved.
- May reduce the illegal dumping of faecal sludge.
- May reduce accidents and spillage.
- Moderate capital and operation costs.
- Pit-emptying teams can receive a payment from the utility/operator per load delivered to the transfer station, thereby ensuring safe disposal of the sludge.
- May encourage more community-level emptying solutions.
- Potential for local job creation and income generation.

Disadvantages

- Fixed stations require expert design, location, and construction.
- May cause blockages and/or disrupt sewage flow.
- Requires an institutional and regulatory framework for taking care of access fees, connection to sewers, or regular emptying and maintenance.
- Can lead to bad odours and vermin if not properly maintained.
- May inconvenience a few for the benefit of the whole community.

Source: (ISF-UTS, SNV, 2016)

CHAPTER 9

Emergency treatment

The primary purpose of emergency treatment is the control the prevention of faecal-related diseases. It is accepted that this may lead to environmental damage (which should be minimized as much as possible), but the health and welfare of affected populations must take priority.

This chapter will concentrate on treatment processes specifically aimed at the acute and stabilization phases of the response. Conventional treatment processes and those that require an extended start-up period will not be covered unless they have a specific relevance these phases.

9.1 Treatment plant rehabilitation

In urban areas, the principal approach should always be to get the existing facilities up and running rather than construct alternative facilities. Chapter 8, section 4 has already discussed the rehabilitation of sewerage systems and pumping stations, but sewage treatment plants should also be reviewed. Many urban communities are at least partly served by a sewerage system, and many sewerage systems discharge to some form of wastewater treatment plant. The damage to a treatment plant may be man-made or due to natural causes; however, in many cases, this only compounds existing damage caused by a long-term lack of investment. There is a wide variety of treatment processes in use around the world, making it impossible to provide detailed advice on what to do in an emergency.

It is highly recommended that the services of an experienced wastewater treatment design engineer are sought to carry out any assessment. After visiting the site and cataloguing the problems, establish an action plan based on the primary objective of emergency relief, which is to:

reduce the exposure of all communities influenced by the treatment plant to the transmission of faecal–oral diseases and disease-bearing vectors.

When developing the action plan, bear the following in mind:

- Involve the local utility responsible for wastewater treatment from the very beginning. They will know much more about the plant than a short visit will discover and will appreciate the merits of possible options. Very often a simple cash injection will enable the utility to carry out the necessary works itself.
- Repair and rehabilitation are generally better than replacement. Rehabilitation fits the knowledge, skills, and equipment availability of local partners. It is also likely to be more sustainable. However, some

plants are so badly designed that they are almost impossible to operate effectively. Assess whether the treatment process is viable, as well as whether it can be rehabilitated.

- Maximize the distance between people and sources of infection. The people at risk may not be the same as those affected directly by the emergency; people living downstream of the treatment plant outfall may be at greater risk than those whom the treatment plant serves. If the treatment plant cannot be rehabilitated and the subsequent pollution contaminates a downstream water source, it may be necessary to consider providing people with an alternative water supply.
- Address the greatest risks to health first. This is probably an element of the treatment process but could be repairs to the security fence around the site.
- Address the problem incrementally and sequentially. Don't try to do it all at once, and bear in mind the capacity of the implementing agency. Start with improvements to the inlet works and gradually work through the treatment process towards the outlet. This will deliver the biggest improvements in wastewater quality quickest. Many downstream processes are dependent on functioning upstream processes. So, for instance, trickling filters will block quickly if primary sedimentation is bypassed. Try to develop a full treatment stream before repairing multiple units of one element.
- Providing even partial treatment is better than no treatment at all. If there is a choice, it is better to partially rehabilitate each treatment element rather than focusing on complete rehabilitation. In that way, all the wastewater receives some treatment before discharge. In the worst case, bypass any treatment processes that are not functioning but be aware of the impact on downstream processes.
- Physical processes such as sedimentation should have priority over chemical and biological processes as they have an immediate positive effect on wastewater quality. Biological processes take time to establish before they begin improving wastewater quality and chemical processes generally require more sophisticated equipment and expensive chemicals that may be harder to source and require a higher skill level to repair and operate.
- The rehabilitation of sludge treatment must go in tandem with wastewater treatment. An effective process for safe containment and treatment is an essential element of any treatment plant rehabilitation. Chemical treatment using calcium hydroxide may have to be used as a temporary measure (see Chapter 9, section 7).

If the plant is beyond immediate repair either:

- establish an uninterrupted wastewater flow through the plant to prevent ponding or overflows to surrounding areas – most treatment plants are

constructed with bypasses for all treatment processes, but, if they are not present, construct a temporary bypass pipe or trench; or
- consider constructing temporary treatment facilities such as waste stabilization ponds or just storage lagoons – many treatment sites have large areas of unused space.

9.2   Coarse filter screens

Coarse filters take many forms but are basically a framework of holes or slots through which sewage or faecal sludge is passed to capture gross solids. In many cases, this is the only pre-treatment applied to faecal sludge, especially where the sludge has been collected manually.

Design
The size of the holes and slots varies but they should be large enough to allow the free flow of sludge or sewage through the filter but small enough to catch any solids that could cause harm to downstream equipment or treatment processes. A slot size of around 6 mm is commonly used as it catches rags, paper, solid anal cleansing materials, and garbage. Screens are normally made of steel rods over a catch pitch (the area below the screen to collect the filtered sludge, its design dependent on the next treatment process), such as that shown in Figure 9.2, but for small plants a simple woven basket may be sufficient. For larger wastewater volumes, a sloping bar screen within a channel may be necessary (Figure 9.1). Screens should preferably be mounted on a slope so that the material caught can be raked to one edge for easy removal. The gross solids collected should be buried.

Advantages
- Generally a simple and quick unit to fabricate or repair using local materials.
- Effective at removing gross solids if regularly cleaned.

Figure 9.1 Coarse inlet screens for faecal sludge treatment

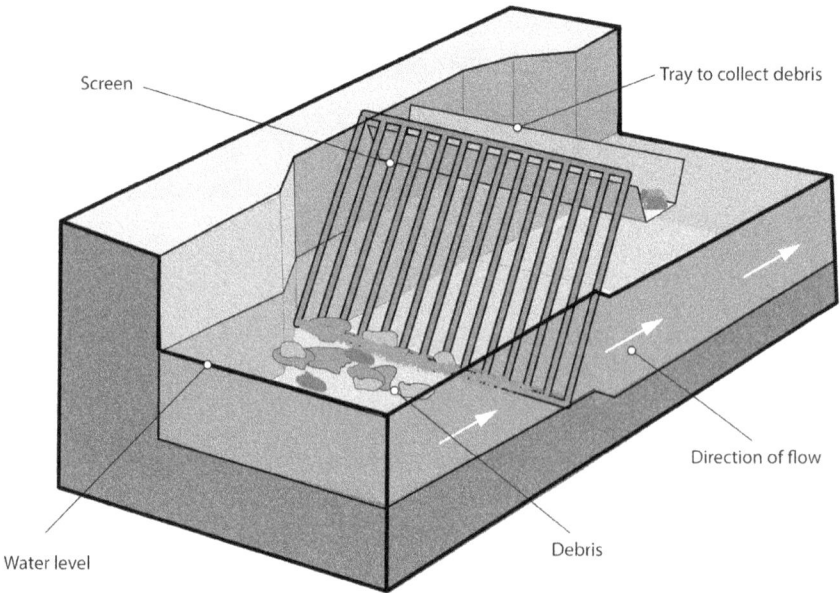

Figure 9.2 Sloping bar screen

Disadvantages
- Blocks quickly if waste contains large amounts of solids.
- Requires constant attention to prevent sludge spillage and screen blocking.

9.3   Grit removal

Both solid and liquid faecal sludge frequently contains large quantities of material of a non-organic origin, such as sand, silt, and general garbage. These can cause major problems with later treatment processes, increasing the sludge accumulation rates, blocking pipes, damaging mechanical equipment, and increasing the thickness of consolidated sludge, thus making it difficult to remove, especially from covered tanks. Most treatment processes that require the separation of solids from liquids will benefit from the removal of non-organic solids beforehand. Both grit removal options require faecal sludge to be liquefied before being treated.

9.3.1 Grease trap

Traps for removing grease from sullage will also effectively remove non-organic materials (see Chapter 14, section 5.2 for further details).

9.3.2 Parabolic grit channels

These are long channels along which the sewage or liquefied faecal sludge flows (Figure 9.3). The speed of flow along the channel is controlled by a weir or a

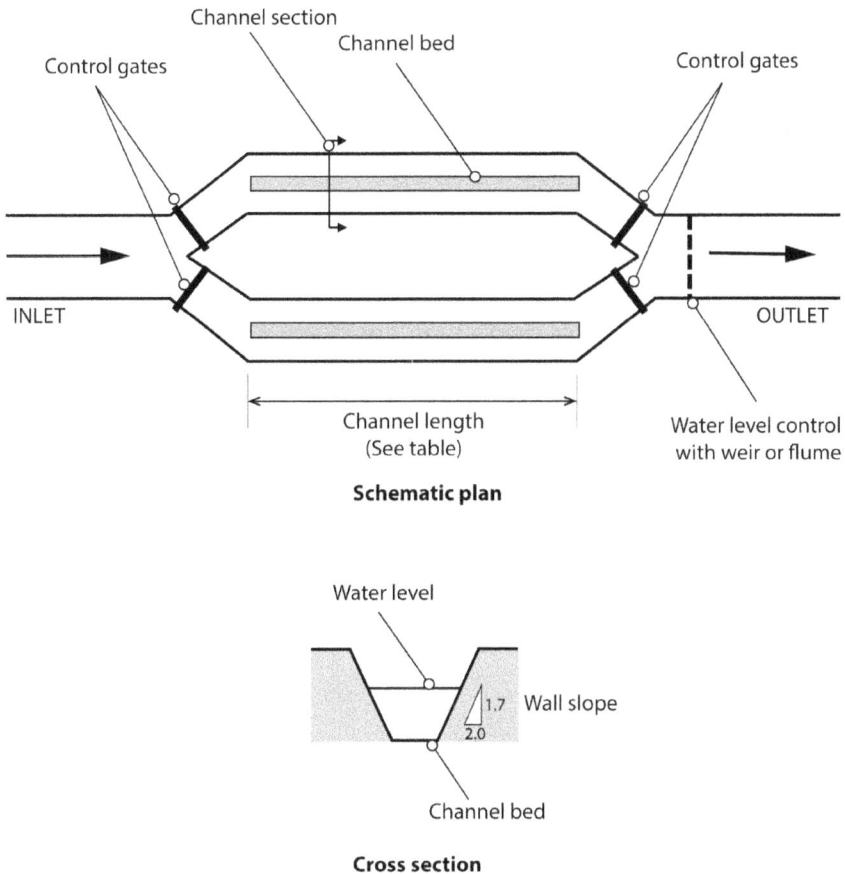

Figure 9.3 Schematic view of simple grit channel
Note: See Table 9.1 for dimensions.

flume so that it is fast enough to carry organic material but allows solid materials to settle to the bottom. The channel should have a parabolic cross-section, but as this is difficult to construct it is usually simplified to a trapezoidal shape. Two channels are used in parallel, each able to carry the total flow. Only one channel is used at a time, the other being emptied and cleaned.

Design
Channels are usually made of concrete or cement-lined brickwork, but in an emergency the channels could be cut out of stable soil and lined with plastic sheeting. Key dimensions for the channel are given in Table 9.1; these are only approximate (for more permeant structures, refer to (** Tayler, 2018). The channels should be straight and horizontal and as smooth as possible. A step down in the outlet channel is useful to prevent water backing up the grit channels. The control gates can be simple pieces of wood tightly fitted

Table 9.1 Dimensions for simple grit channel

Flow (l/s)	Water depth (mm)	Liquid surface width (mm)	Channel bed width (mm)	Channel length (m)
10	113	442	309	3.4
20	175	570	364	5.3
30	227	662	395	6.8
50	313	798	430	9.4

Source: adapted from (** Tayler, 2018)

against the channel wall and bed. When a channel requires cleaning, the control gates at both ends of the grit channel should be closed and the standing water within removed manually or with a pump. The grit can then be dug out and disposed of, usually by burial.

Advantages
- Simple design and construction.
- Easy to clean.
- Few moving parts.

Disadvantages
- Require a large amount of space.

9.4   Batch sedimentation

Some treatment processes require the separation of suspended solids from the liquid. For small treatment plants, open-top prefabricated storage tanks can be adapted for this process (Figure 9.4). The tank is filled with screened faecal

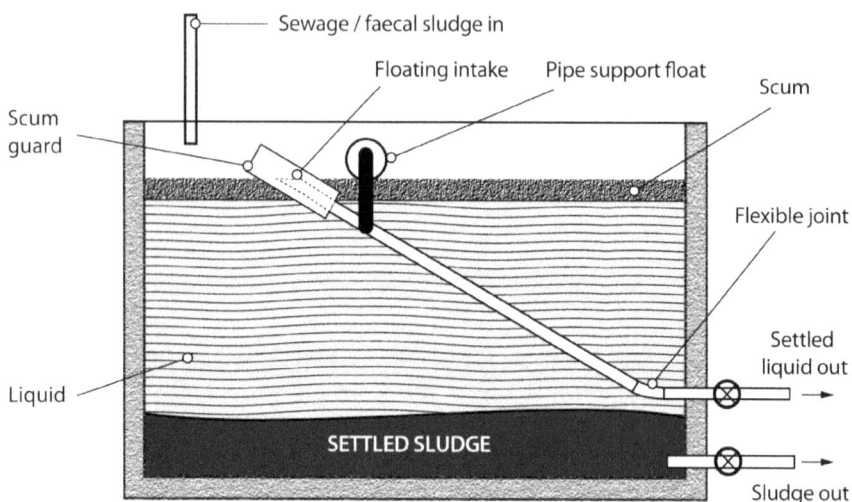

Figure 9.4 Batch settlement tank

Box 9.1 Batch settlement tank design for 10,000 population

Assuming fresh faecal sludge is being collected in a low-income country where water is used for anal cleaning.

The daily sludge production rate is assumed as: 2.6 + 0.25 = 2.85 litres/person/day Table 4.2)

Daily total sludge production is 2.28 × 10,000/1,000 = 28.5 m³/day

Assume the use of standard prefabricated steel tanks with butyl liners of 24 m³ (3.66 m diameter × 2.29 m high).

The full capacity of the tank cannot be used for settlement as a 0.3 m gap must be left between the maximum liquid level and the tank top to prevent disturbance from the wind. Space should also be allowed for sludge accumulation if the sludge cannot be removed at the end of each cycle. Assume 0.3 m.

Effective capacity of the tank is (π × 3.66²/4) × (2.29–0.3–0.3) = 17.8 m³

A single tank can receive around 60 per cent of a day's demand, therefore four tanks are required. This would also provide sufficient additional volume for variations in the volume of faecal sludge delivered to the site.

Operation

Two tanks are required for a batch system, each tank being used on a two-day cycle.

Day 1: The first tank is filled followed by the second one. They are allowed to stand overnight allowing adequate time for settlement.

Day 2: The third and fourth tanks are filled sequentially while tanks one and two are being emptied and cleaned.

Day 3: Tanks one and two are sequentially filled again while tanks three and four are emptied and cleaned.

In theory, the plant could function with two tanks, the first tank being emptied around two hours after it was filled and the second tank emptied the following morning for reuse in the afternoon, but this would require additional personnel and close supervision. There would also be no flexibility in case of breakdowns or sudden surges in sludge volume.

---

sludge which is then allowed to stand undisturbed for a period. Suspended solids will either settle to the bottom (sludge) or float to the surface (scum). Once settlement is complete, the settled liquid is drawn off through a floating intake arm suspended just below the scum layer. Once all the liquid has been drawn off, the sludge and scum can be removed separately. At least two tanks are required so that one tank can be emptied and refilled while the other is standing full.

Design

A settlement time of around 1–2 hours is normally sufficient, but carry out a settlement test to confirm (Appendix 5). Sludge removal can be difficult if the bottom of the tank is horizontal; mechanical assistance such as the use of a pump is normally required. Scum can be removed by carefully raking it to the side of the tank where it is removed manually with a bucket. Box 9.1 gives a design example.

Advantages
- No start-up period.
- Quick to install if storage tanks and fittings are readily available

- Flexible design – a number of smaller tanks can be used if larger ones are not available.
- Flexible sizing – tanks can be readily added or subtracted to meet demand.
- Tanks can be of any shape (in plan).

Disadvantages
- System operation involves significant personnel and close management.
- High risks of spillage.
- Only part of a treatment process – requires earlier and later elements to match the settlement process.
- Unsuitable for faecal sludge with high solids content, such as pit toilet sludge.

9.5 Horizontal flow settlement ponds

In some cases, a continuous settlement system is more appropriate than a batch system. In general, a tank is fabricated by excavating a shallow pond. The spoil can be used to construct embankments around the pit to increase the depth of the pit. Normally, the base and sides of the tank are lined with an impermeable membrane (in the short term, plastic sheeting will do) to protect the groundwater and prevent short-circuiting of the liquor (Figure 9.5). Where groundwater contamination is not a problem, it may be possible to leave the pond unlined, particularly where the soil has a high clay content. Settlement tanks are also used to increase the solids content of sludge prior to application on drying beds.

Design
Designing and constructing large ponds is a specialist activity. It is strongly recommended that someone experienced in ground excavation is involved.

Hydraulic design details are shown in Table 9.2. A baffle board is fixed across the full width of the tank close to the inlet and outlet to help spread the flow evenly across the width of the tank and to prevent floating solids overflowing into the outfall. The board should extend 0.3 m above and about 0.5 m below the water level. The invert of the outlet pipe fixes the liquid level in the tank and the inlet pipe invert should be above the top water level.

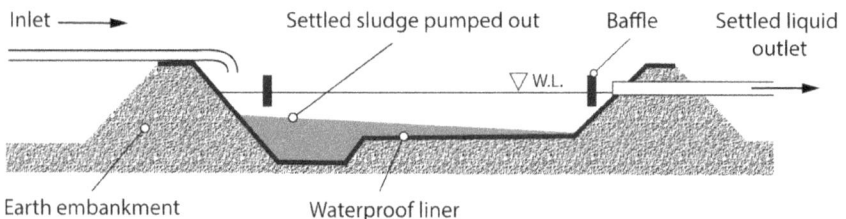

Figure 9.5 Horizontal flow settlement tank

Table 9.2 Design criteria for horizontal flow settlement ponds

Criteria	Value
Tank retention time for liquid effluent	2–3 hours
Peak factor	2–4
Horizontal velocity	<0.005 m/s
Surface loading (the volume of liquid passing through the volume of the tank below 1 m² of tank surface area in a day)	24–48 m³/m²/day (use lower number for smaller tanks)
Liquid depth	1.5–3 m^
Width to length ratio	1:3–1:4
Freeboard	0.3 m
Desludging	Daily

Note: ^The bottom of the tank near the inlet can be deeper if required as that is where most sludge will settle.

Where the tank is to be filled by manual carriage or tanker, a dedicated inlet chamber should be constructed rather than throwing the sewage directly into the pond. A design example is given in Box 9.2. Constructing two tanks is preferable; it provides spare capacity should supply increase and allows one tank to be taken out of commission while desludging.

Box 9.2 Horizontal settlement pond for 10,000 population

Assuming fresh faecal sludge is being collected in a low-income country where water is used for anal cleaning.
The daily sludge production rate is assumed as: 2.6 + 0.25 = 2.85 litres/person/day (Table 4.2)
Daily total sludge production is 28.5 m³/day
Equivalent to an average flow 1.19 m³/hr.
Peak factor depends on how the influent is delivered, assume 4.
Assume a faecal sludge retention period of two hours. This is normally sufficient, but check with a settlement test (Appendix 5).
Tank settlement volume = average hourly flow x peak factor x retention time = 9.52 m³
Assume surface loading rate (SLR) = 24 m³/m²/day (Table 9.2).
Therefore, tank surface area = daily design flow x peak factor/SLR = 28.5 × 4/24 = 4.75 m²
Therefore, liquid depth = 9.52/4.75 = 2.0 m
Tank length to width ratio = 3:1 (Table 9.2)
Therefore, tank width = $(4.75/3)^{1/2}$ = 1.25 m
Tank length = 3.75 m
Check horizontal velocity = average flow × peak factor/cross-section area = [(1.19 × 4)/(3600)]/(1.25 × 2.0) = 0.00053 m/s : OK <0.005 (Table 9.2)

Sludge storage
Assume sludge generation of = 0.25 litres/person/day) (Table 9.2).
Assuming emptying daily (recommended) and stored in the pit below the base of the tank.
Volume of sludge = 0.25 × 10,000/1000 = 2.5 m³
Assume the pit is the width of the tank and 2.0 m long.
Pit depth = sludge volume/pit area = 2.5/(2 × 1.25) = 1.0 m

(Continued)

Box 9.2 Continued

Total tank depth
Freeboard = 0.3 m to prevent high winds disturbing the scum layer
Scum thickness should be minimal if the tank is regularly desludged.
Total tank depth (excluding a sludge well) of 2.0 + 0.3 = 2.3 m
Note: Excavated tanks usually have sloping sides. The dimensions calculated here should correspond to the mid height of the pond. Suggested embankment slopes are given in Appendix 3.

Source: Adapted from (The Institution of Water Pollution Control, 1980) and (Kamel, n.d.)

The tank should be desludged when the sludge level reaches the maximum design thickness. In general, it is better to have a small sludge-holding capacity than a large one as this will encourage frequent desludging. Long periods between desludging will lead to sludge digestion that will cause bad odours and the resuspension of sludge. Also, desludging tends to be forgotten when there are long gaps! Desludging is usually undertaken using a mechanized pump

Advantages
- Requires less personnel than a batch system to operate and maintain.
- Uses less space than a batch system.
- Converts an irregular supply feed to a more consistent flow for downstream treatment processes.
- Accommodates rapid changes in inlet flow rates.

Disadvantages
- Requires more on-site engineering skills for construct.
- Unsuitable for faecal sludge with high solids content, such as pit toilet waste.
- Unsuitable for ground with a high water table or rock close to the surface.

9.6 Unplanted drying beds

Unplanted drying beds separate the liquid and solid components of the faecal sludge for further treatment or disposal (Figure 9.6). The top surface of the bed is usually a single-sized sand that is underlain with layers of gradually increasing aggregate size, contained within low walls. Sometimes the top of the bed is covered with a permeable membrane to simplify sludge removal. Larger particles are trapped by the top sand layer and the finer and liquid portions percolate through the bed to the filter base, where they are channelled to an outlet for further treatment or to evaporate. Once the sludge has partially dried it can be manually removed (usually with a small layer of sand) and taken for further treatment.

Where groundwater conditions are appropriate, the bottom of the filter bed may be left unsealed and the filtrate allowed to infiltrate directly into the subsoil. This should be a very temporary solution and not used until a full assessment of the immediate and long-term impact on groundwater and nearby surface quality is carried out.

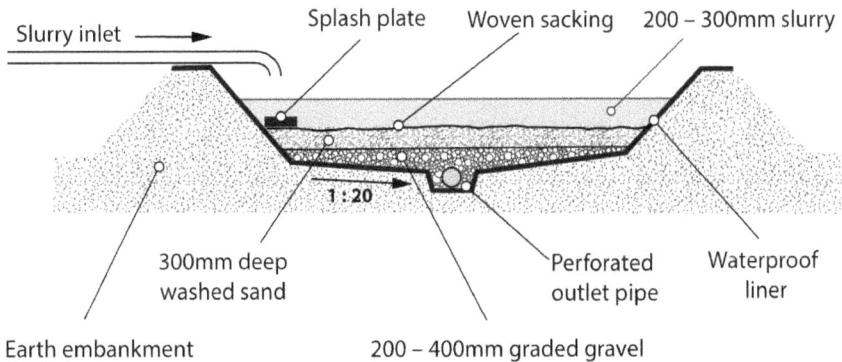

Figure 9.6 Sludge drying bed

Design

The number of drying beds required depends on the volume of faecal sludge to be treated and the time taken to fill the bed, dry the sludge, and remove the dried cake (sludge). Wet sludge is discharged onto a bed to a depth of 200–300 mm. It is then left until the sludge has dried sufficiently to allow its removal. The inlet should be equipped with a splash plate to prevent erosion of the sand layer and to allow for even distribution of the sludge. Even in dry areas provision should be made for covering the beds in case of rain. Consider sludge-drying beds where the following conditions are met:

- The material to be treated has a solids content between 3 and 14 per cent. This is difficult to measure in the field, so samples should be sent to a local public health laboratory for testing. Alternatively, construct a small pilot unit (say 1.0 m²) and see what happens when the faecal sludge is applied at the correct dosage.
- The volume to be treated is low. Most existing treatment plants are designed for less than 20 m³/day, but it should be possible to consider up to 50 m³/day.
- Land is available.
- The management capacity, knowledge, and skills required for more complex treatment processes are lacking.

Loading and resting periods should not exceed four to five weeks, although much longer cycles are common. Beyond that, the quality of the filtrate may start decreasing, while sludge does not thicken further. Shorter periods, even as little as one week, can also be used where local circumstances allow. For an optimal design, it is recommended to test the settling capacity of the sludge beforehand. Drying beds are normally designed on the peak monthly flow. Table 9.3 shows typical design criteria for drying beds and Box 9.3 a design example. Thickening the sludge (such as in a settlement tank) prior to application to the drying beds will considerably reduce the volume to be dried.

Table 9.3 Drying bed design criteria

Parameter	Range	Notes
Sand effective size	0.3–0.75 mm	200–300 mm thick
		Wash sand before use
		Uniformity coefficient <3.5
Gravel sizes	Bottom 20–40 mm	200–450 mm thick
	Middle 5–155 mm	Sizes can be varied to suit local availability
	Top 2–6 mm	
Wet sludge loading depth	200–300 mm	
Dewatering time	4–15 days (hot/arid climate with covered beds)	Guide figures to achieve 15–30 per cent solids
	15–30 days (temperate/wet climate with covered beds)	Actual times dependent on local conditions
		Longer drying times will result in higher solids content
Solids loading rate	50 kg total solids/m²/year (temperate climates)	Not generally used in initial design
	1–200 kg total solids/m²/year (tropical climates)	

Source: Adapted from (** Tayler, 2018) and (** Gensch, et al., 2018)

Box 9.3 Drying beds design for a population of 10,000

Assuming fresh faecal sludge is being collected in a low-income country where water is used for anal cleaning.

  The daily sludge production rate is assumed as: 2.6 + 0.25 = 2.85 litres/person/day (Table 4.2).
  Daily total sludge production = 28.5 m³/day
  Weekly production = daily production × 7 = approx. 20.0 litres/person/week
  Therefore, weekly total is 200 m³
Assuming a six-day collection cycle, then daily collection is 200/6 = 33.3 m³ (figures rounded up).
Assume hydraulic loading depth on beds is 200 mm.
Assume a new bed is used every day (this can be changed if the bed sizes turn out to be too large or too small).
  Area of a single bed is 33.3/0.2 = 166.5 m²
  Bed shape can be anything to suit the site layout but assume rectangular with a side ratio 2:1, therefore side lengths are 9.1 m × 18.2 m
  This is a large bed – in practice, it is likely to be divided up into a number of smaller beds for ease of operation.
Assume, based on local experience, the drying time for sludge is nine days (this must be determined locally).
  Total drying cycle is loading time + drying time + emptying time
  If it takes two days to empty, then total drying time is 12 days, one of which is a non-working day.

(Continued)

Box 9.3 Continued

The number of drying beds required is 11 (see figure below) but add an extra bed to allow for maintenance and repair.

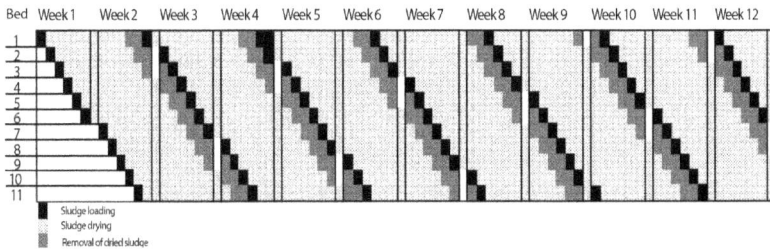

To check the solids loading rate to see if it falls within the acceptable range, we have to know the total solids of the influent.

From Table 4.2 assume the quantity of dry faecal matter is 38 g/person/day out of a total volume of 2.85 litres of faecal sludge, giving a total solids of 38/2.85 = 13.3 g/litre.

The figure above shows that six full cycles of the drying beds' use takes 11 weeks, so the number of bed-loading cycles in a year is 6 × (52/11) = 28.36 (round to 28)

Hydraulic loading rate = hydraulic loading rate (0.2) × total solids loading rate (13.3) × number of cycles per year (28) = 74.5 kg total solids/m²/year

This is quite low for a hot, arid climate (Table 9.3). The process would benefit from a pre-thickening process that would considerably reduce the volume of liquor to be dried and hence the area of drying beds required.

Note: Excavated tanks usually have sloping sides. The dimensions calculated here should correspond to the mid-height of the pond. Suggested embankment slopes are given in Appendix 3.

Source: Adapted from (** Tayler, 2018)

Advantages

- Good dewatering efficiency, especially in hot dry climates.
- Simple technology using local materials and skills.
- Low capital cost (excluding land).
- Simple operation.
- Suitable for a high water table, hard ground, and flooded environments.
- Can partially treat excreta from storage vault-based toilets (such as chemical toilets, storage tank toilets, or septic tanks).

Disadvantages

- Requires a large land area.
- Public health risks whenever excreta are handled, transported, or processed.
- Nuisance from odour and flies.
- Labour intensive, requiring constant supervision.
- Limited sludge stabilization and pathogen reduction.

9.7 Chemical treatment using calcium hydroxide

As the name suggests this treatment process is based on the application of chemicals to disinfect faecal sludge so that it is safe (from a health perspective) to discharge into the local environment. A range of chemicals have been tried but the most commonly used in emergencies is calcium hydroxide – $Ca(OH)_2$ – often known as 'hydrated lime'. When thoroughly mixed with faecal sludge slurry it raises the alkalinity to such an extent that the faecal bacteria are destroyed. The process poses significant environmental risks and should only be seen as a temporary emergency response.

Fresh and pre-stored faecal sludge is first passed through a coarse filter to remove gross solids (see Chapter 9, section 2). It then enters a tank where it is thoroughly mixed with hydrated lime, usually in liquid form (Box 9.4), until the pH of the mixture has been raised to and remains at more than 12 for at least two hours (Box 9.5). The hydrated lime has to come into direct contact with pathogens to destroy them, so it is essential that the sludge contains no large solids and that mixing is thorough. A schematic view of the whole process is shown in Figure 9.7.

Box 9.4 Mixing calcium hydroxide

- To liquefy hydrated lime, mix in the ratio 1 kg of dry powder with 3–4 litres of clean water. Place the required quantity of water in a clean container. Add the dry powder and stir vigorously for a few minutes until the powder is fully mixed with the water. Use the mixture as soon as possible after mixing. If the liquid is not to be used immediately, cover the container with a tight-fitting lid and briefly stir the contents before use.
- The quantity of hydrated lime to add is normally in the ratio 0.25–0.35 kg of dry hydrated lime to 1 kg of dry sludge solids.

Note: Measuring the dry solids component of faecal sludge requires access to a laboratory where a fixed weight of fresh sludge can be dried to determine the proportion of dry solids in the sludge. In the absence of laboratory facilities, use trial and error to determine the quantity of hydrated lime necessary to achieve a pH of 12 for two hours. An approximation of the dry solids component can be found in Table 4.2

Box 9.5 Mixing hydrated lime and sludge

- To be effective, the hydrated lime must come into direct contact with individual particles of the sludge. In most cases this means that hand mixing will not work.
- Gross solids must be removed from the sludge first and then extensive mechanical mixing used to ensure a full contact between the two elements.
- For emergency settings, this is best achieved by using a batch mixing system and a mechanical mixer such as the one shown in Figure 9.8. The container must be relatively small to ensure thorough mixing of all the contents.
- The mixing process should always be carried out in a well-ventilated area, preferably outside. This is because the gases and aerosols generated are a significant health risk.

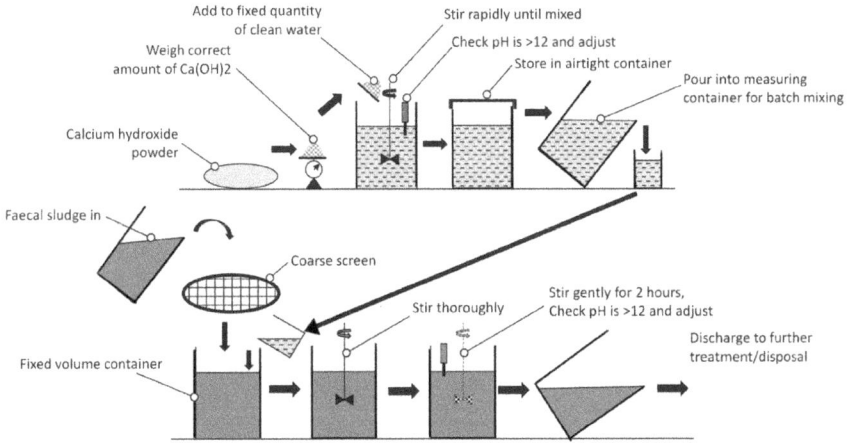

Figure 9.7 Schematic view of treatment process using calcium hydroxide

Figure 9.8 Hand held mechanical mixer

Design

The amount of hydrated lime required is largely dependent on the quality of the hydrated lime and the quantity of solids in the sludge. A design example is given in Box 9.6. After two hours of mixing, most pathogenic bacteria will have been destroyed and the treated mixture should meet WHO guideline limits of $<10^3$ cfu/100 ml. The process does not destroy all helminths and there is no information on its effect on viruses. After the disinfection process is complete, the pH gradually declines to a more stable level and the sludge can be disposed of in its mixed state; however, it is usually separated into its liquid and solid components, each receiving further treatment prior to disposal or reuse. Details of the design of sludge drying beds is given in Chapter 9, section 6.

Box 9.6 Design of a calcium hydroxide treatment process for 10,000 people

Assuming fresh faecal sludge is being collected in a low-income country where water is used for anal cleaning.

   The daily faecal sludge production rate is assumed as: 2.6 + 0.25 = 2.85 litres/person/day (Table 4.2)

   Daily total is 28.5 m³/day

   Weekly production is approx. 20.0 litres/person/week

   Therefore, weekly total is 200 m³

   Assume a six-day working week and faecal sludge delivered to the treatment plant over a 10-hour period.

   Daily faecal sludge delivered to plant is 33.3 m³.

   Hourly delivery rate is 3.3 m³.

   From Table 4.2 assume that the quantity of dry faecal matter is 38 g/person/day out of a total volume of 2.85 litres of faecal sludge, giving a total solids of 38/2.85 = 13.3 g/litre (same as kg/m³).

   Therefore, weight of faecal solids arriving per hour is 13.3 × 3.3 = 43.9 kg.

   Assume an application rate of 0.3 kg of dry hydrated lime per kg of dry sludge (Box 9.4) gives an hourly hydrated lime use of 0.3 × 43.9 = 13.17 kg

   Assuming a dilution of 3 litres of water to 1 kg hydrated lime, the volume of water required per hour is 39.6 litres.

Notes: In practice it would be better to carry out pilot tests on the calcium hydroxide slurry concentration and the mixing rate with the faecal sludge to obtain the best mix. This calculation will give you only an approximation of the quantities required.

   The choice of batch size will depend on the space available and the size of containers available. A new batch could be started every hour or only once or twice a day. This is a large volume of faecal sludge to treat using this method at a single site and great care would be needed to protect the health and safety of the operators with such large quantities of calcium hydroxide being used.

BEWARE!

Hydrated lime can irritate the skin, eyes, lungs, and digestive system.
Direct or indirect contact with faecal sludge is associated with microbial and chemical hazards. Various pathogenic helminth (worms), protozoa (amoeba), oocysts, bacteria, and viruses can cause a wide range of sometimes serious diseases.

Workers must be fully trained in the handling of hydrated lime and faecal sludge, be provided with the full complement of personal protective equipment (PPE), and be aware of the correct course of action in case of an emergency.

Advantages
- Kills most pathogenic bacteria.
- Can be set up rapidly using local materials, equipment, and skills.
- Works immediately.
- Alters the structure of the final effluent, improving dewatering efficiency.
- Minimal power requirements.
- Simple to operate, maintain, and monitor.

- Reduces odours and vectors.
- Uses widely available and relatively cheap chemicals.
- Space requirement less than for other treatment options.

Disadvantages
- Does not destroy all helminth eggs.
- Requires very close monitoring of day-to-day operation to ensure effective pathogen reduction.
- A hazardous chemical requiring careful handling.
- Space must be well ventilated and operatives must wear full PPE.
- Calcium loses its effectiveness if stored for extended periods or in poor conditions.
- The final effluent has a high organic load (Biological Oxygen Demand [BOD], Suspended Solids [SS], nitrogen, etc.) unless it undergoes additional treatment.
- Highly variable quality of locally sourced calcium hydroxide.

Section adapted from Reed & de Pooter, 2018.

9.8   Waste stabilization ponds

Waste stabilization ponds (WSPs) are large, artificial ponds that separate the solids and liquid portions of faecal sludge and use anaerobic and aerobic micro- and macro-organisms to digest organic material, eliminate pathogens, and stabilize the effluent, making it suitable for safe discharge to the environment. Most treatment plants consist of a series of three ponds: anaerobic, facultative, and aerobic (maturation) (Figure 9.10). Anaerobic and facultative ponds are designed for BOD removal, while aerobic ponds are designed for pathogen removal. In the early stages of an emergency, it is common to start by constructing a single pond. An anaerobic pond is most suitable for fresh faecal sludge and a facultative pond for partially treated sludge such as from pit toilets, drying beds, ABRs, and septic tanks.

The anaerobic pond is the primary treatment stage and reduces the organic load in wastewater through solids separation and organic decomposition within the settled sludge and the suspended liquor. Anaerobic bacteria convert organic carbon into methane and, through this process, remove up to 60 per cent of BOD (Figure 9.9). The facultative pond further reduces the BOD. The top layer of the pond receives oxygen from the air through wind mixing and algae-driven photosynthesis. The lower layer is deprived of oxygen and becomes anaerobic. Any remaining solids that can settle are digested at the bottom of the pond. Aerobic and anaerobic organisms work together to achieve a BOD reduction of up to 75 per cent. The final maturation pond is aerobic and designed to polish or finish the effluent, including removing pathogens. It is the shallowest pond, ensuring that sunlight penetrates the full depth for photosynthesis to occur; this, in conjunction with wind action, increases the oxygen content of the effluent.

Figure 9.9 Anaerobic waste stabilization pond

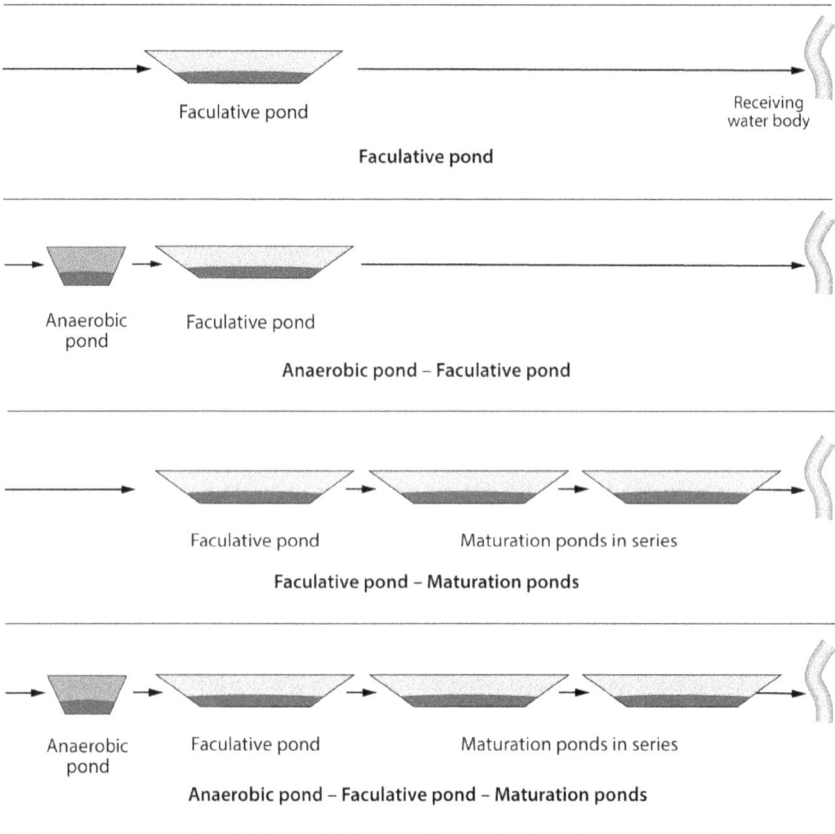

Faculative pond

Receiving
water body

Faculative pond

Anaerobic
pond

Faculative pond

Anaerobic pond – Faculative pond

Faculative pond

Maturation ponds in series

Faculative pond – Maturation ponds

Anaerobic
pond

Faculative pond

Maturation ponds in series

Anaerobic pond – Faculative pond – Maturation ponds

Figure 9.10 Waste stabilization ponds systems
Source: Adapted from (Sperling, n.d.)

Table 9.4 Design data for emergency waste stabilization ponds

Pond type	Dimensions	Notes
Anaerobic ponds	• 2–5 m deep. • 1–7 days' retention time. • Rectangular, with length to breadth ratio of less than 3.	Efficiency improved by the installation of mechanical aerators
Facultative ponds	• 1–2.5 m deep. • 5–30 days' retention time. • Rectangular, with length to breadth ration of more than 10.	Efficiency improved by the installation of mechanical aerators
Maturation ponds	• 0.5–1.5 m deep. • 3–5 days' retention time. • Rectangular, with length to breadth ratio of more than 10.	Additional ponds can be added to improve pathogen removal

Design

A summary of the key design data for emergency ponds is given in Table 9.4. A good hydraulic design is also important to avoid short-circuiting: i.e. wastewater travelling directly from inlet to outlet. The inlet and outlet should be as far apart as possible, and baffles can be installed near the inlet and outlet to ensure complete mixing within the ponds and to avoid stagnating areas. Pre-treatment is recommended to prevent scum formation and to hinder excess solids and garbage from entering the ponds. To protect ponds from runoff and erosion, a protective berm or mound should be constructed around each pond using excavated material. See Appendix 3 for details of embankment design.

The hydraulic design of waste stabilization ponds is complex and not normally a field-based calculation. However, a simple (and conservative) design calculation is given in Box 9.7.

Advantages
- Treats human wastes in a low-tech way with very few moving parts, machines, or energy requirements.
- Withstands sudden shock loading from high flows or increased organic material.
- High reduction of solids, BOD, and pathogens.
- Low operating costs.

Disadvantages
- Requires large land area for ponds so often some distance from the communities served.
- Transporting excreta to the treatment site can be complex and expensive.

Box 9.7 Design of waste stabilization ponds for a population of 10,000

Assuming fresh faecal sludge and some bathing, laundry, and cooking waste is being collected in tankers in a low-income community where water is used for anal cleaning. The daily wastewater production rate is assumed as: 12.85 litres/person/day (adapted from Table 4.2).
   Total daily flow = 128.5 m^3
   Anaerobic pond
   Assume a retention time of two days (longer in colder climates), therefore total tank volume = 128.5 × 2 = 257 m^3.
   Assume a tank depth of 2.0 m, therefore tank surface area = 129 m^2.
   Assume rectangular tank with length to breadth ratio of 2, therefore surface dimensions are 8 m × 16 m.
   Facultative pond
   Assume a retention time of five days (longer in colder climates), therefore total tank volume = 642.5 m^3.
   Assume a tank depth of 1.5 m, therefore tank surface area = 428.3 m^2.
   Assume rectangular tank with length to breadth ratio of 10, therefore surface dimensions are 6.5 m × 65 m.
   Maturation pond
   Assume a retention time of four days (longer in colder climates), therefore total tank volume = 514 m^3.
   Assume a tank depth of 1.0 m, therefore tank surface area = 514 m^2.
   Assume rectangular tank with length to breadth ratio of 10, therefore surface dimensions are 7.2 m × 72 m.
   Note: Excavated tanks usually have sloping sides. The dimensions calculated here should correspond to the mid-height of the pond. Suggested embankment slopes are given in Appendix 3.

Source: Adapted from (Kayombo, n. d.)

- Complex calculations for comprehensive design.
- Unsuitable in very cold climates.
- Difficult to remove sludge on the rare occasions necessary.
- Anaerobic ponds can emit strong odours.

9.9   Septic tanks

A septic tank is a water-tight container that receives excreta from water-flushed toilets and grey water from bathing and washing. They routinely have two compartments divided by a perforated wall (Figure 9.11) that reduces hydraulic disturbance and increases settlement efficiency. The unit has an overflow that discharges the treated effluent for further treatment or disposal. Bad odours and other gases from the septic tank discharge either through the vent pipes in the inlet sewer or through vents fitted to the septic tank cover and are prevented from causing a problem in the toilet by the water seal within the toilet interface. Vent pipes should always discharge above head height. Treatment in septic tanks occurs through a combination of settlement, floatation, and sludge digestion. In addition, grease and oils float to the surface to form a layer of scum. Over

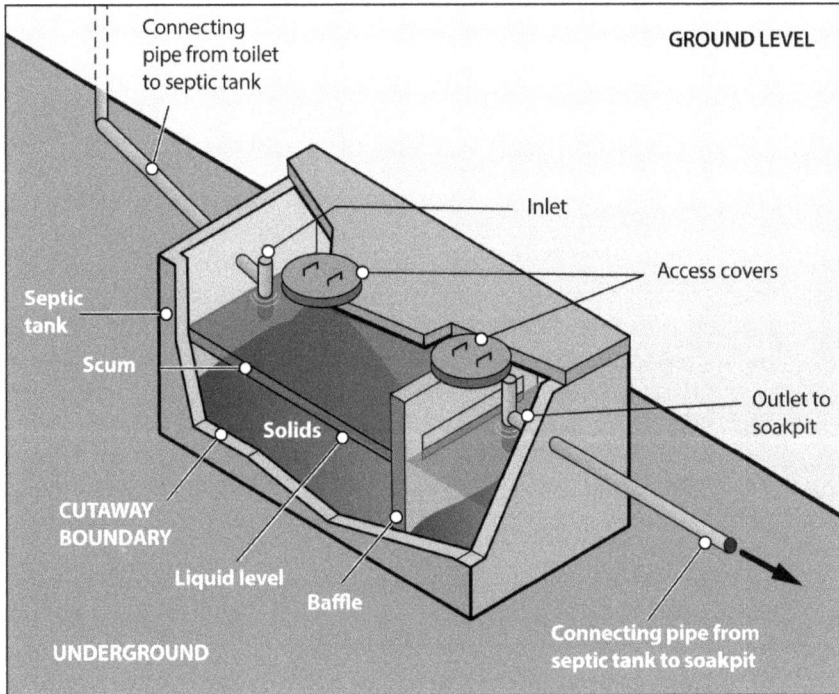

Figure 9.11 Septic tank

time, the sludge at the bottom of the tank is compressed by the weight of new material settling on top, increasing its density.

Septic tanks are commonly used in emergencies either because they existed prior to the emergency or they are constructed to treat faecal and other wastes from institutions or communal toilet blocks. In most cases, septic tanks are constructed on-site, usually from a combination of bricks or blocks and concrete. However, in some countries prefabricated septic tanks are available, although their effectiveness in emergency settings has not been positive.

A well-designed septic tank can remove around 50 per cent of suspended solids, 30–40 per cent of BOD, and 70–80 per cent of E. coli, although efficiencies vary greatly depending on operation, maintenance, and climatic conditions. Septic tanks in very cold climates are very inefficient. In general, the larger the wastewater flow and the bigger the septic tank, the more efficient the treatment process. The effluent should always receive additional treatment before being considered safe. Options for this are given in Chapter 10.

9.9.1 Tank components

In small tanks, the base is usually made of unreinforced concrete about 100–150 mm thick. This is thick enough to withstand the uplift pressure when the tank is empty and act as a foundation for the walls. In larger tanks, the floor will have to be reinforced to carry the imposed loads and prevent cracking. Walls are made of bricks, blocks, stone, or concrete. The walls must be watertight if there is a potential for contaminating a water source; if there is no potential, it is not critical. The space between the walls and the sides of the excavation should be filled with a granular material such as fine gravel. The tank cover is usually made of concrete. Access holes with covers should be provided over the inlet, outlet, and dividing walls to allow for desludging and maintenance. In large tanks, the cover may be made of a series of removable slabs that are light enough so that each can be removed manually for desludging.

As with all settlement tanks, the construction of the inlet and outlet is critical to the performance. The wastewater must enter and leave the tank with the minimum of disturbance and the tank must be easy to maintain. For smaller units, a 'tee' piece design is the most appropriate. In larger units, the outlet tee piece should be replaced by a weir and scum board plate (Figure 9.12). The scum board is to prevent the floating solids from being washed into the outlet pipe. Do not use a weir on the inlet, as it will become fouled with large solids.

The dividing wall must have a facility for allowing the effluent to pass from one compartment to another. This can be achieved either by installing tee pieces through the wall at water level or by leaving holes (slots) in the wall. Table 9.5 suggests the minimum number of holes to leave in the dividing wall for different flow rates. In all cases, the openings should be designed to minimize turbulence by ensuring that the average velocity through them does not exceed 0.1 m/s. The openings must be in the middle of the settlement zone, not in the sludge or scum zones.

9.9.2 Design

Many different formulae have been developed for designing septic tanks, and they give wildly differing dimensions. Some countries have their own design

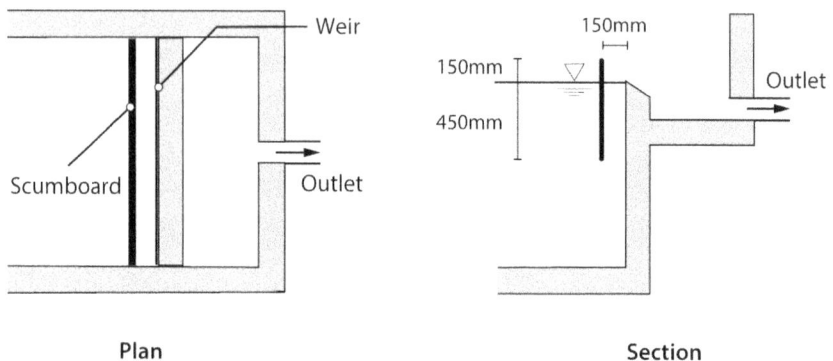

Plan   Section

Figure 9.12 Alternative septic tank outlet design

Table 9.5 Rectangular septic tank internal dimensions

Wastewater flow (m³/d)	1	2	3	5	8	11	16	22
Liquid, sludge, and scum depth (m)	1.2	1.3	1.4	1.8	2.0	2.1	2.3	2.5
Total tank depth (m)	1.7	1.8	1.9	2.3	2.5	2.6	2.8	3.0
Tank width (m)	0.8	1.1	1.2	1.5	1.7	1.9	2.1	2.3
1st compartment length (m)	1.7	2.2	2.4	3.1	3.5	3.7	4.1	4.6
2nd compartment length (m)	0.8	1.1	1.2	1.5	1.7	1.9	2.1	2.3
Number of 10 cm × 10 cm holes in dividing wall	4	4	4	7	11	15	22	30

Note: These dimensions assume the septic tank is desludged at least once a year.
Source: (** Harvey, 2007)

codes and these should be followed if working in that country. For ease of construction, the base of the tank is usually constructed horizontally, but in larger tanks the base often slopes towards the inlet end of the tank. This is because most of the sludge collects under the inlet. In the absence of local guidelines, the data given in Table 9.5 is recommended. Box 9.8 contains a specimen design calculation using these recommendations. In areas of high groundwater or that are subject to flooding, empty septic tanks could float if not properly designed (see Appendix 6 for more details).

9.9.3 Location

The low water use in emergencies creates difficulties in transporting suspended solids long distances through pipes (see Chapter 8, section 4 for more details).

Box 9.8 Designing a septic tank for a communal toilet block

Design a septic tank for the toilet block shown in Figure 9.13
Calculate the daily volume of wastewater
Assume the worst-case scenario for estimating the population using the toilet block.
Number of users per cubicle = 50 (Table 5.2)
Total population served = 50 × 4 = 200
Assume the per capita volume of excreta and anal cleaning water to be 2.85 l (Table 5.2)
Assume daily quantity of water used daily for toilet flushing is 2 litres per person
Assume daily quantity of water used for bathing and laundry is 5 litres per person
Total daily wastewater generated per person is 9.85 litres – say 10 litres
Total volume of wastewater generated per day is 10 x 200 ÷ 1,000 = **2 m³**
Calculate septic tank dimensions using Table 9.5
Minimum internal tank depth (liquid + sludge + scum +ventilation) = 1.8 m
Tank width = 1.1 m
Length of first compartment = 2.2 m
Length of second compartment = 1.1 m
Number of 10 cm × 10 cm slots in the dividing wall = 4

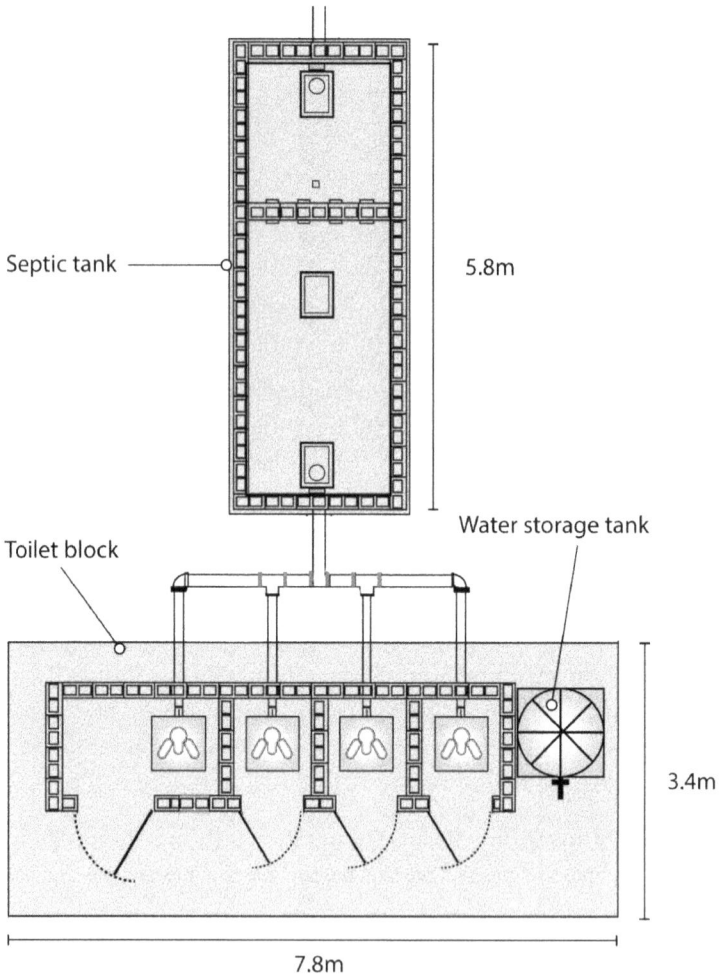

Figure 9.13 Toilet block close coupled to a septic tank
Source: Adapted from (Kayombo, n. d.)

Septic tanks must be placed as close as possible to the toilets, bathrooms, etc. that they service, as shown in Figure 9.13.

9.9.4 Alternative designs

When flow rates are very low, the septic tank can be replaced by individual vaults constructed directly below or adjacent to the defecation point. The tank can be fitted with an overflow pipe discharging to additional treatment processes, as shown in Chapter 10. Alternatively, it may be more cost-effective and quicker to construct the tank out of upturned concrete culverts, as shown in Figure 9.14.

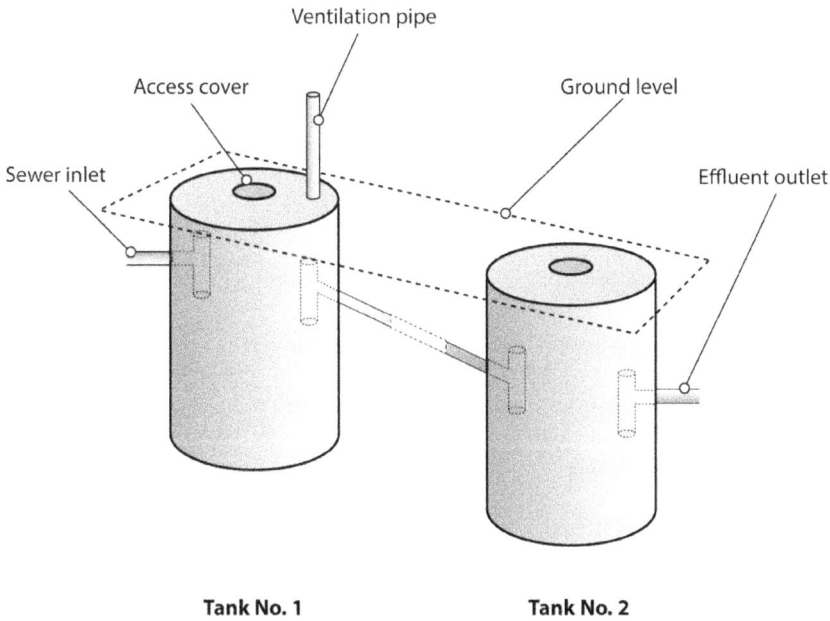

Figure 9.14 Small septic tank using concrete culverts

The sizes of these units vary, largely depending on the size of culverts available locally, but typically each tank is around 0.8–1.0 m in diameter and 1.0 m high. The first tank has a watertight base and is also covered with a removable concrete slab and ventilation pipe. The second tank is sometimes sealed but may be fitted with a granular filter in the base if it is safe to infiltrate the effluent into the subsoil. These tanks are not as efficient as rectangular tanks but are easy, quick, and relatively cheap to construct. They may be a good solution for the short to medium term while water use remains low.

9.9.5 Operation and maintenance

Only fully degradable toilet paper or water should be added to toilet waste to prevent blockages. Large amounts of chemicals (including bleach) should not be flushed into the tank as it will kill the bacteria that digest the organic content. Sludge should be removed at regular intervals (annually if the dimensions shown in Table 9.5 are used). Suggested emptying options are given in Chapter 7. A small amount of sludge should be left in the tank after emptying to seed the digestion process.

Large communal septic tanks in refugee settings are usually managed by centralized maintenance crews. Smaller household or shared septic tanks are typically the responsibility of the users for operation, maintenance, and cleaning, although support is typically given for desludging.

Advantages
- Suited to refugee settings where water is used for anal cleansing and toilet flushing.
- Low operating costs and long service life.
- Well-known technology, commonly with local infrastructure for construction and emptying.

Disadvantages
- High construction costs compared with simple pit-based systems.
- Requires relatively large volumes of water to function efficiently.
- Requires periodic desludging.
- Low reduction in pathogens and organics.
- Effluent and sludge require additional treatment.
- Tanks are frequently difficult to access with sludge-emptying machinery.

9.10   Up-flow anaerobic sludge blanket reactor (UASB)

UASBs separate wastewater into three phases: sludge, liquid effluent, and gas. The wastewater to be treated is introduced at the bottom of the tank and rises through a suspended sludge blanket. Anaerobic bacteria in the sludge blanket break down organic material in the influent, converting it into biogas which rises up through the reactor. Baffles separate the gas from the liquid flow and direct it into one or more gas hoods, from which it is drawn off and either used or flared. The effluent overflows at the weirs and the solids remain in the blanket or settle. Together, the gas hood and baffle arrangement is referred to as the GLS (gas–liquid–solids) separator. Good contact between sludge and wastewater is achieved through even distribution of the inflow across the bottom of the UASB and through the agitation caused through the production of biogas. Figure 9.15 shows a typical UASB reactor (** Tayler, 2018).

Design
To facilitate sludge withdrawal, a series of valves should be provided at intervals of about 50 cm across the height of the sludge blanket, with the first valve positioned 15–20 cm above the reactor floor. Most UASBs have volumes in the range of 1,500–3,000 m³, giving a capacity range of 6,000–12,000 m³ per day for a six-hour retention period. UASBs have been constructed with volumes as low as 65 m³, giving a capacity of the order of 260 m³ per day, but these have usually been pilots for larger installations.

Operation and maintenance
UASBs require operators who have a basic understanding of the processes that take place in the reactor and who know and follow the practices required to ensure good reactor performance. Successful operation of UASBs depends on regular monitoring of sludge levels and suspended solids concentrations and withdrawal of excess sludge from the reactor.

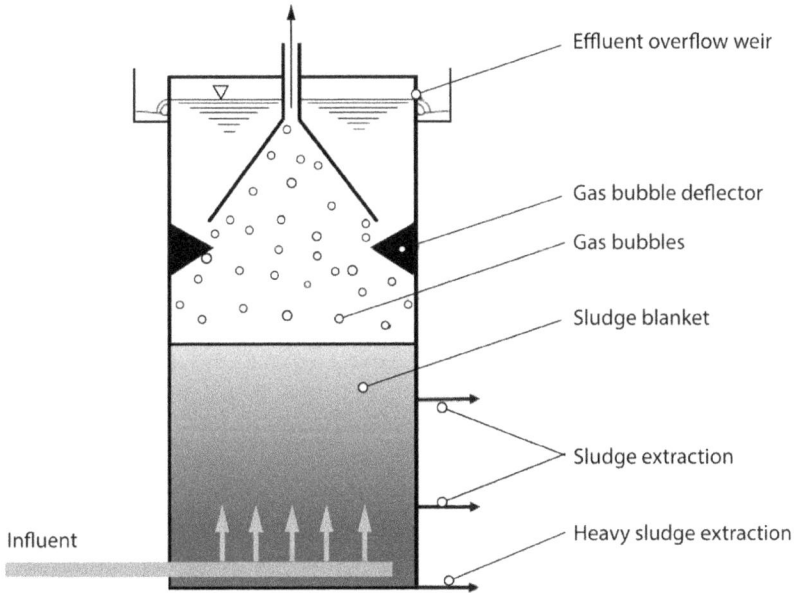

Figure 9.15 Up-flow anaerobic sludge blanket reactor

For more detailed information on design and operation refer to (van Lier, 2010).

Advantages
- Most appropriate for highly diluted wastewater with low concentrations of organic matter.

Disadvantages
- Impossible to maintain a sludge blanket with influent flows varying from peak flows during the day to zero at night.
- Anaerobic baffle reactors (see Chapter 9, section 12) generally work better than UASBs.
- Very sensitive to good operator management.
- Existing plants frequently found not to be working effectively.
- Low reduction in pathogens and organic matter.
- Extended start-up time.
- Unsuitable for early stages of an emergency response.

9.11 Biogas chambers

If human excreta is stored under anaerobic conditions (preferably combined with agricultural wastes) and mixed with water, it will give off gas as it decomposes (Figure 9.16). Given the right temperature and mix of wastes,

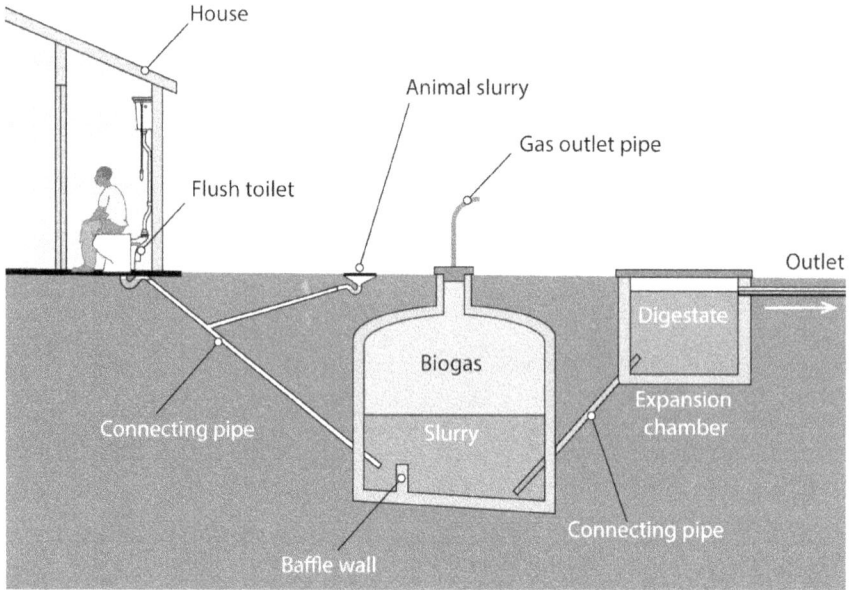

Figure 9.16 Small biogas toilet

much of the gas will be methane, which is flammable. The mix of gases produced is called biogas. Biogas generation is widely used for digesting bulk volumes of excreta-based sludge, such as at sewage works, but it also has been incorporated into domestic toilets in both stable and emergency settings, with mixed success.

Design, construction, and operation
The unit consists of a large watertight domed tank, often buried in the ground. A 100 mm diameter pipe is installed between the bottom of the tank and the outlet from the toilet defecation point. Branches from that pipe go to other waste sources such as other toilets or inlets for farm manure. The toilet interface and all other inlet points should be fitted with a water seal. A similar pipe is installed between the bottom of the tank on the opposite side to the inlet pipe and a collection chamber. The top of the tank is fitted with a gas-tight seal into which a 50–75 mm plastic pipe is installed, capped by a shut-off valve.

There are no specific design parameters for the size of the tank, but it should hold wastes for around 30 days. A tank of around 1.2–1.6 m³ should be sufficient for serving a single household owning a few cattle or sheep or a group of around six families. The position of the collecting chamber controls the liquid level in the biogas tank. The bottom of the chamber should be located so that the tank is about two-thirds to three-quarters full when first filled. The plastic pipe in the top of the chamber is the gas draw-off point. It should be connected via a gas scrubber to the heat-using appliances.

To initiate the system, the biogas tank is filled with a mixture of water and animal manure until it starts to overflow into the collecting chamber. Users can then start to defecate in the toilet. Excreta must be flushed into the inlet pipe with about 5 litres of water. When the ambient temperature is above 20°C, the organic matter in the biogas tank will begin to anaerobically digest, giving off gas. The gas rises to the top of the tank where it is stored until required.

The digested slurry in the collecting chamber is unsafe. The anaerobic digestion process does not kill all pathogens and the short holding time is not enough for them to die naturally. The slurry should be buried or disposed of in another safe manner.

Traditionally, biogas plants have been very expensive to build, but there are now prefabricated units available that reduce cost and speed up construction (two to three days). The main problem is with operation and maintenance, particularly adding non-biodegradable material or fibrous matter such as straw. This leads to loss of gas production and blockage of the digester tank. They are appropriate only in communities with a commitment to recycling organic wastes and little access to alternative power sources. Care should be taken to prevent tank flotation, especially in areas subject to a high water table or flooding. More advice on this topic is given in Appendix 6.

Where only excreta are added, the volume of gas produced may not be sufficient for daily needs. Sharing the plant between multiple families may be more appropriate as it reduces the unit cost and the plant works more effectively. However, if too many families share, there will be social difficulties with sharing the gas produced. Around six families appears to be a reasonable number.

Advantages
- Produces biogas that can be used for cooking and lighting.
- Suitable for those who use water for anal cleaning.
- Toilet cubicle can be used for bathing, laundry, etc.

Disadvantages
- Expensive and complex to install.
- Requires expertise during the commissioning phase.
- Users must be educated in the details of operation.
- Unsuitable for users who clean themselves with non-biodegradable materials.
- Only viable where alternative low-cost fuel sources are not available.
- Families must be willing to equitably share the gas produced.
- The process ceases to work well at ambient temperatures below 20°C and completely below 5°C.
- Very difficult to repair if blocked through inappropriate use.
- Low reduction in pathogens and organic matter.

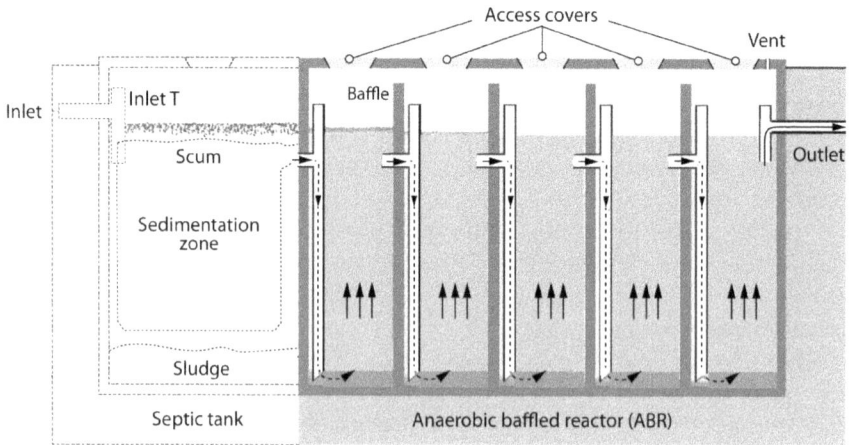

Figure 9.17 Anaerobic baffle reactor
Source: Adapted from (** Gensch, et al., 2018)

9.12   Anaerobic baffle reactor

An anaerobic baffled reactor (ABR) is a single-compartment septic tank connected to a second tank containing a series of vertical baffles (Figure 9.17). Settled wastewater is forced to flow under and over the baffles from the inlet to the outlet with the up-flow chambers providing enhanced removal and digestion of organic matter. ABRs can treat raw, primary, and secondary treated sewage, as well as grey water (with organic load).

Based on very limited data from full-scale ABRs, the BOD may be reduced by up to 90 per cent, which is far superior to its removal in a conventional septic tank. However, its removal of faecal coliform is around the same at approximately 70 per cent, and therefore additional treatment is required to produce a safe effluent.

9.12.1 Design

There are no currently accepted guidelines for the design of ABRs. In most cases to date, pilot plants have been constructed to determine local characteristics prior to full-scale construction. This is patently unrealistic for emergency settings, so Table 9.6 provides suggested key design factors for use in emergencies and Box 9.9 provides a worked example.

9.12.2 Applicability

An ABR for 20 households would take several weeks to construct. If reinforced fibre plastic ABR prefabricated modules are used, the time required for construction is much less (three to four days). Once in operation, three to six months (up to nine in colder climates) is needed for the biological environment to become established and maximum treatment efficiency

Table 9.6 Suggested data for designing ABRs in emergencies

Design criteria	Value
Wastewater retention time (after exiting the septic tank)	48–72 hours
Peak up-flow velocity	0.6–1.0 m/hour
Number of chambers	4–8
Width of individual chamber	Between 0.75 m and 0.5 × liquid depth
Liquid depth	1.8–2.5 m

Source: Adapted from (** Tayler, 2018)

Box 9.9 Design of an anaerobic baffle reactor

Using the case study shown in Box 9.8, design an ABR to replace the second chamber in the septic tank.

Assume
Retention time of 48 hours
Peak up-flow velocity of 1.0 m/hour

From septic tank design
Daily design flow is 2 m³
Liquid depth is 1.3 m
Tank width is 1.1 m

Calculations
Total liquid volume = 2 × 2 = 4.0 m³
Assume the width of the tank and the liquid depth are the same as the septic tank, i.e. 1.1 m and 1.3 m respectively.
Length of tank = 4/(1.1 × 1.3) = 2.8 m
Minimum cross-sectional area (A) of a single baffle chamber (i.e. the area between two baffle plates) = Q/V, where Q is the peak flow in m³/hour and V is the peak flow velocity of 1.0 m/hour
Average flow per hour = 2/24 = 0.083 m³/hour
Allow for flow not being uniform throughout the day by multiplying by a peak factor of 3. Therefore, peak flow = 0.083 × 3 = 0.25 m³/hour
Therefore A = 0.25/1 = 0.25 m²
Distance between baffles = A/tank width = 0.25/1.1 = 0.23 m
Number of chambers in the ABR = total tank length/distance between baffles = 2.8/0.23 = 12
More information of the design of ABRs, including a detailed design calculation, is given in (** Tayler, 2018).

---

to be reached. ABRs are thus not appropriate for the acute response phase and are more suitable for the stabilization and recovery phases. ABRs are most suited to neighbourhood systems but can also be used at household or catchment levels. They are also suitable for public buildings such as schools. Even though ABRs are designed to be watertight, it is not recommended to construct them in areas with high groundwater tables or where there is frequent flooding because of the possibility of flotation (see Appendix 6). Alternatively, prefabricated modules can be placed above ground. ABRs can be installed in every type of climate, although the efficiency will be lower in colder climates.

9.12.3 Operation and maintenance

ABRs are relatively simple to operate. Once the system is fully functioning, specific operational tasks are not required. To reduce start-up time, the ABR can be inoculated with anaerobic bacteria, for example by adding septic tank sludge or cow manure. The system should be checked monthly for solid waste, and the sludge level should be monitored every six months. Desludging is required every two to four years, depending on the accumulation of sludge at the bottom of the chambers, which reduces treatment efficiency. Desludging is best done by mechanized pumping, for example by a vacuum pump and tanker. A small amount of sludge should be left to ensure that the biological process continues.

Advantages
- Resistant to organic and hydraulic shock loads.
- No electrical energy is required.
- Low operating costs.
- Long service life.
- High reduction of BOD.
- Low sludge production, and the sludge is stabilized.
- Moderate area requirement (can be built underground).
- Simple to operate.

Disadvantages
- Long start-up phase.
- Requires expert design and construction.
- Low reduction of pathogens and nutrients.
- Effluent and sludge require further treatment and/or appropriate discharge.
- Needs strategy for faecal sludge management (effluent quality rapidly deteriorates if sludge is not removed regularly).
- Needs water to flush.
- Clear design guidelines are not available yet.

This section largely adapted from Tayler, 2018.

9.13 Planted drying beds

A planted drying bed is similar to an unplanted drying bed (see Chapter 9, section 6), but it has the added benefit of transpiration and enhanced sludge treatment due to the plants in the bed (Figure 9.18). The key benefit is that fresh sludge can be applied directly to the bed and needs to be removed only every three to five years.

Design

The beds dewater, stabilize, and reduce the volume of the sludge while evaporation and transpiration reduce effluent quantity. Plant root systems maintain filter porosity by creating pathways through the thickening sludge that allow water to percolate easily. Ventilation pipes connected to the drainage system contribute to aerobic conditions in the filter. Sludge is applied on a batch system then allowed to dry out prior to the next application. Multiple beds are therefore required for a continuous system. The sludge is applied to the surface and the filtrate flows down through the subsurface where it is collected in drains.

Careful management is required to ensure that plants do not dry out and die, and to maintain plant density at an acceptable level. In dry climates, beds should be watered regularly to avoid plant death.

Most of the general construction design details are shown in Figure 9.18 but additional information is given in Table 9.7 and a design example in Box 9.10. Further information on excavation is given in Appendix 3.

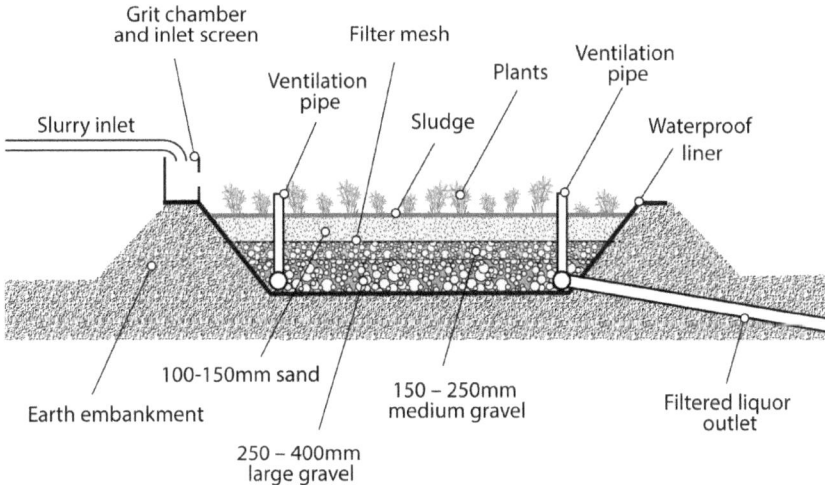

Figure 9.18 Planted drying bed

Box 9.10 Planted drying beds design for a population of 10,000 (Adapted from (** Tayler, 2018))

Assuming fresh faecal sludge is being collected in a low-income community in a hot climate where water is used for anal cleaning.
The daily sludge production rate is assumed as: 2.6 + 0.25 = 2.85 litres/person/day (Table 4.2)
Weekly production is approximately 20.0 litres/person
Daily total is 28.5 m³
Weekly total is 200 m³
Assuming a six-day collection cycle, then daily collection is 33.3 m³ (figures rounded up)

(Continued)

Box 9.10 Continued

From Table 4.2 assume the quantity of dry faecal matter is 38 g/person/day out of a total volume of 2.85 litres of faecal sludge, giving a total solids of 38/2.85 = 13.3 g/litre (kg/m³)
Total solids loading (kg/year) = 28.5 × 13.3 × 365 = 138,353 kg/year
Assuming a 200 kg/m² maximum solids loading rate
Surface area required = 138,353/200 = 690 m² (rounded)
Assuming effluent loading to each bed once a week, then six beds will be required (or multiples of six if bed sizes are too large).
Daily bed area required = 690/6 = 115 m²
This is equivalent to six beds of 7.6 m × 15.2 m.
Beds don't have to be perfectly rectangular, provided they can be loaded with sludge uniformly and are free draining.

Source: Adapted from (** Tayler, 2018).

Advantages
- Appropriate for urban or camp settings that generate a constant slurry supply.
- Functions well in humid climates where technologies relying on evaporation work poorly.
- Can handle high loadings.
- Better sludge treatment than unplanted drying beds.
- Construction using local skills and materials.
- Manpower required for removing plant growth much less than for desludging unplanted drying beds.
- Possible income from sales of plants or plant products.
- Reduced exposure to health risks for operating staff.

Disadvantages
- Beds can take many months to mature before they can accept their design loading.
- Both the sludge and the effluent require further treatment to remove all pathogens.
- Requires a large land area.
- Staff skilled in plant management essential.
- Often problems with odour and flies.

9.14   Horizontal flow constructed wetlands

Constructed wetlands are engineered wetlands designed to filter and treat different types of wastewater, mimicking processes found naturally. They are typically used to treat septic tank or waste stabilization pond effluent that has already been partially treated before it is discharged into the environment. They are rarely an appropriate technology for the earlier stages of an emergency because of their long start-up time and their complex operation and design. A short introduction is given here so that readers are aware of them should they come up in discussions.

Table 9.7 Design data for planted drying beds

Data	Value	Comments
Filter material		
Base filter layer (stones)	200–400 mm diameter	
	250–400 mm thick	
Upper gravel layer	5–15 mm diameter	
	150–250 mm thick	
Sand	2–6 mm diameter	
	100–150 mm thick	
Well-draining soil	50 mm thick	Not always present
Free space above filter	1.0 m	
Plant types	Reeds (Phragmites sp.), antelope grass (Echinochloa sp.), and papyrus (Cyperus papyrus)	Local, non-invasive species can also be used if they grow in damp soil conditions, are resistant to salty water, and readily reproduce after cutting
Slurry application rates	100–200 kg total solids/ m²/year	Warm climates
	50–70 kg total solids/m²/ year	Cold climates
Sludge depth	70–100 mm thick, with 3–10 days drying time between dosage	

Source: Adapted from (** Tayler, 2018) and (** Gensch, et al., 2018).

They are typically constructed in a large gravel and sand-filled channel that is planted with aquatic vegetation. A number of variations are available with different flow arrangements. However, typically, as the wastewater flows through the channel, the gravel and sand filter out particles and pathogen removal is accomplished by natural decay, predation by higher organisms, and sedimentation (Figure 9.19).

The water level is generally kept to at least 15 cm below the surface of the gravel and sand to prevent overland short-circuiting and the bed is typically lined with clay or plastic sheeting to prevent leaching into the environment. Any aquatic wetland plants with large root systems can be used. Communal or centralized wetlands are usually managed by centralized maintenance crews. The effluent requires additional treatment to meet minimum health standards. Overall, facultative ponds will almost always offer a better option for simple secondary treatment than constructed wetlands.

Design

The impermeable sides and base of the wetland structure contain a gravel bed, typically 30–60 cm deep and planted with wetland plants. The length-to-width

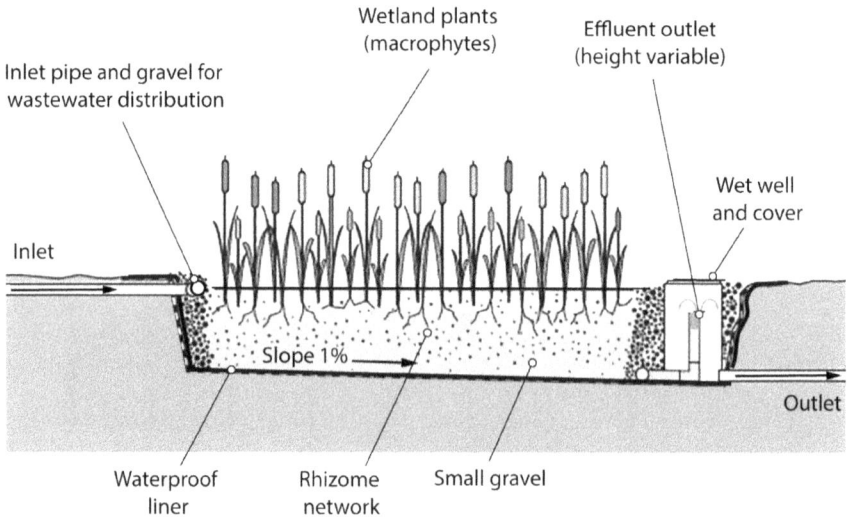

Figure 9.19 Horizontal flow constructed wetland
Source: Adapted from (** Gensch, et al., 2018)

ratio should be at least 2:1, sufficient to create a long travel path for the effluent and to reduce the likelihood of short-circuiting. Wastewater enters at one end of wetland 'cells' and must be distributed across the full width of the wetland. It then flows through the gravel and exits the cell at the other end. As it does so, a combination of physical and biological processes removes the suspended solids, organic carbon, and nitrogen loads in the wastewater. The wetland plants must have deep roots and should be able to adapt to humid environments with slightly saline and nutrient-rich soil conditions. *Phragmites australis* or *communis* (reeds) are often chosen because they form a matrix of rhizomes efficient at maintaining the permeability necessary for large filtration and they also decrease the risk of clogging.

Loading rates for domestic sewage are about 30 g $BOD5/m^2$/day d at 25°C, which is below the rate that can be achieved in a facultative waste stabilization pond at the same temperature.

Advantages
- Robust performance.
- Less susceptible to input variations than septic tanks provided the flow regime is properly regulated.
- Effectively treats domestic primary or secondary treated sewage, as well as grey water.
- Provides tertiary treatment of effluents from septic tanks in a low-tech way in situations where there is tight control over environmental discharges such as high water tables, flooded environments, or urban settings.
- Feasible mid- to long-term solution.

Disadvantages
- All wastewater must undergo preliminary treatment to remove suspended solids prior to application.
- Unsuitable for high organic and suspended solids loading such as septic tank and pit toilet sludge.
- Long start-up period while plants become established.
- Unsuitable for the acute response phase.
- Takes up large areas of land and needs careful monitoring to ensure they are functioning correctly.
- Complex design and very little experience of use with faecal sludge.

9.15   Package treatment units

Many of the materials used in faecal sludge management are prefabricated: pipes, floor slabs, vehicles, etc. are all pre-manufactured components that contribute to the treatment process. To date, however, there is no widely accepted and used complete pre-packaged unit for the treatment of faecal sludge in emergencies. Many companies around the world design, fabricate, and install wastewater treatment plant packages in settled communities. They are commonly designed for domestic or industrial wastewater and very few companies manufacture plants for treating high-strength faecal sludge. The plants are not prefabricated but manufactured to order, and so there is a large time lag between ordering and commissioning, making them suitable only for the recovery phase of emergencies and for long-term use.

For many years, attempts have been made to develop an emergency treatment plant package but, as yet, they have not been successful because of the challenges mentioned above. Work is continuing to develop treatment packages, but they are unlikely to be suitable for an acute response because of delivery difficulties. However, they may be appropriate for the stabilization and recovery phases. Furthermore, the design and operation of many of these plants are proprietary and they must be maintained and serviced by the company that manufactures them.

Design
With such a wide range of products produced worldwide, it is not possible to provide useful and consistent design data.

Advantages
- Complete package capable of the full treatment of faecal wastes from start to finish.
- Usually provided with dedicated maintenance and operational support and/or local staff training.
- Feasible mid- to long-term solution.

Disadvantages

- Long lead time from ordering to fully functioning plant.
- Complex design, generally tailored to specific site needs.
- Minimal (successful) experience of treatment of high-strength faecal sludge in emergency settings.
- Very expensive.
- Consistent and reliable power supply essential.
- Skilled staff essential for sustainable operation and maintenance.

CHAPTER 10
Tertiary treatment, disposal, and reuse

10.1   Infiltration

Many treatment plant failures can be traced back to the failure of the effluent disposal system. This is particularly true of systems that infiltrate effluent into the ground. That is why this section covers infiltration systems in such detail. It isn't as simple as many think! Subsurface infiltration is suitable for effluent with relatively low suspended solids, such as the overflow from well-designed and managed septic tanks, settlement tanks, drying beds, anaerobic baffle reactors, unplanted drying beds, planted drying beds, constructed wetlands, and sullage from water points. A range of infiltration systems are discussed below.

Before beginning your design, two key questions should be addressed:

- Will the effluent contaminate a drinking water source (see Chapter 4, section 4)?
- Is the substrate (soil/rock profile immediately around and below the infiltrating surface) suitable for infiltration (Table 4.3)?

10.1.1 Infiltration pits

Infiltration pits are large holes in the ground that store effluent. They are commonly between 2 m and 5 m deep and 1.0–2.5 m in diameter. Their size below the inlet pipe is dependent on the infiltration capacity of the side walls and on having a volume larger than the total daily effluent flow. Effluent entering the pit soaks into the surrounding soil through the sides of the pit; the pit base is ignored because it quickly becomes blocked. The pit can be lined or filled with large stones, blocks, bricks, etc. (Figure 10.1). The fill supports the pit walls and cover but does not play any part in the treatment process. The volume taken up by the fill should be deducted from the volume of the pit when checking whether the pit volume is large enough. Any lining must be porous so that the wastewater can reach the soil interface. However, the top 0.5 m of the pit (measured from ground level) must have a sealed lining to support the pit cover and keep out surface water and vermin. An example of infiltration pit design is given in Box 10.1.

Soak pits are generally suitable only for disposing of small quantities of wastewater, such as from a single home, small institution, or individual tap/handpump apron. Flows from groups of houses or large institutions would require a pit too large to be economical. The infiltration process is primarily

Clay or similar fill in space
between lining and soil

Air-tight access cover

Inlet

Top 0.5m of lining sealed
with cement mortar

Gravel fill in space
between lining and soil

Liquids percolate
into the soil

Open jointed lining

Lined infiltration pit

Top 0.5m sealed lining

Inlet

Boulders used
to support pit walls

Liquids percolate
into the soil

Unlined infiltration pit

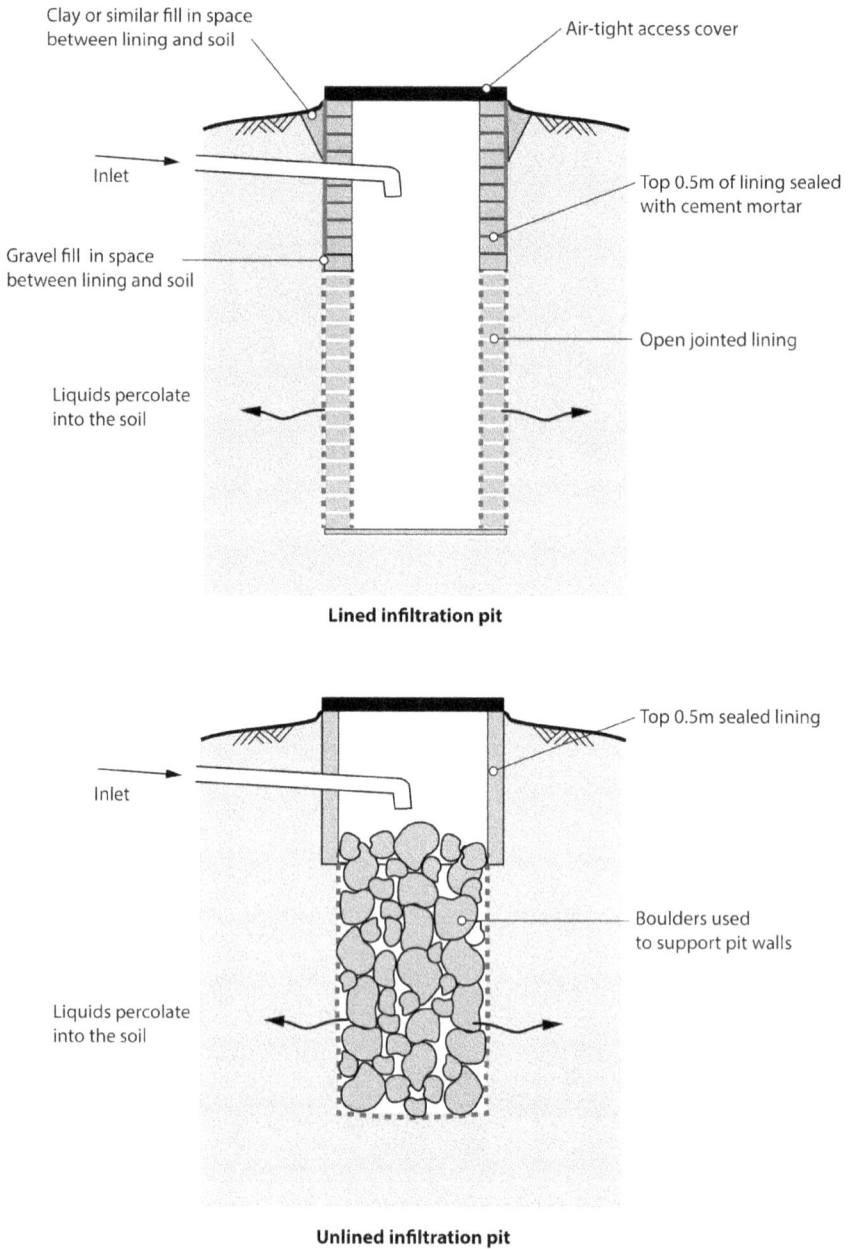

Figure 10.1 Lined and unlined infiltration pits

aerobic, and therefore it is preferable to keep the pit above the groundwater table. If the groundwater is potable, the bottom of the pit should be at least 1.5 m above the water table.

Box 10.1 Infiltration pit design and case study

Theory

Assume: A = pit wall area below the inlet pipe available for infiltration (m²)

Q = daily wastewater flow (litres/day)

S = soil infiltration rate (litres/m²/day) (from Table 4.3) – measure infiltration rates at different depths and take average

D = assumed pit diameter (m)

C = pit circumference (m)

H = pit depth below inlet pipe (m)

V = pit volume (m³)

Then: Calculate soil infiltration area (A = Q/S)

Assume a pit diameter and calculate the circumference (C = π × D)

Calculate the pit depth below inlet (H = A/C)

Calculate the total pit depth by adding the distance from the ground level to the bottom of the inlet pipe or 0.5 m, whichever is the greatest

Check that pit liquid volume below inlet (excluding fill material) is greater than the daily influent flow. (Q ≤V)

If the volume is too small, increase the depth or pit diameter until sufficient volume is achieved

Square and rectangular pits

The design process is the same except that:

C = length of the pit perimeter = ([length + width] × 2)

H = pit depth below inlet pipe = A/C

Appendix 3 gives more information on safe excavation methods.

Case study

Design an infiltration pit to dispose of the effluent from the septic tank designed in Box 9.8.

Assume the soil is sandy loam and the water table is at a depth of 5 m but is highly saline and unsuitable for a drinking water supply.

Daily effluent flow rate = 2.0 m³

Soil infiltration rate = 25 litres/m²/day (Table 4.3)

Soil infiltration area = 2.0 × 1,000/25 = 80 m²

Assume a pit diameter of 1.5 m (diameter at soil interface not inside lining)

Pit circumference = π × 1.5 = 4.71 m

Depth of pit required is = 80/4.71 = 17 m

This is much too deep for a single pit; it would also require most of it to be constructed below the water table. The options are to dig multiple pits (i.e. six pits 3 m deep) or use a larger diameter pit. The choice would depend on the relative costs and the space available. In any case, infiltration pits are probably not the best solution for this problem.

Advantages

- Simple technology widely used.
- Single pit systems require very little space.
- Uses local skills and materials.
- Particularly common for systems serving a small number of families.

Disadvantages

- The way that infiltration pits work is widely misunderstood, leading to poor design and construction.

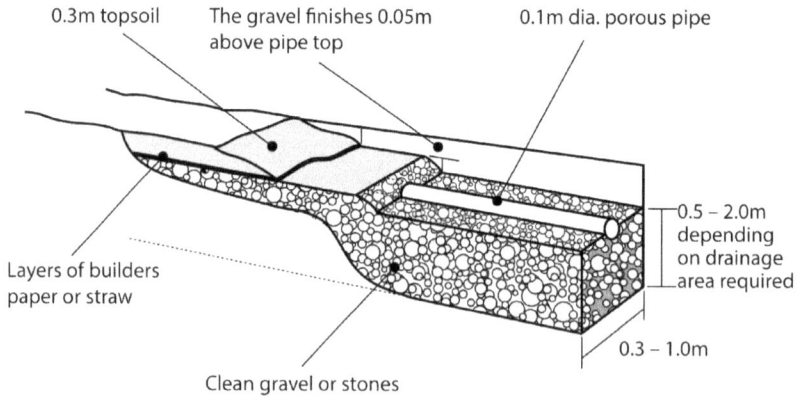

Figure 10.2 Section through an infiltration trench

- Potential for localized groundwater contamination, especially in high-density housing areas where shallow groundwater is used for drinking.
- Pits that use stones to support the side walls and roof are very difficult to clean if a poorly maintained septic tank leads to sludge overflow and blockage.

10.1.2 Trench infiltration

Disposing of effluent in a trench provides a higher surface area for the volume of soil excavated. It also utilizes the upper soil layers, which tend to be more porous. Pipes disperse the effluent along a series of trenches that have been filled with coarse gravel (Figure 10.2). The pipes are porous (Box 10.2) so that the effluent can seep out into the surrounding gravel. They are laid horizontally to spread the effluent evenly along the whole length of the pipe. The gravel disperses the effluent from the pipe to the walls of the trench, where it is absorbed into the surrounding soil. The trench bed is ignored as it quickly blocks. A layer of paper, straw, or porous plastic sheeting covers the top of the pipe. This allows air to enter the trench and gases to escape but prevents the topsoil from mixing with the gravel and blocking the trench. Key design characteristics are given in Table 10.1 and Box 10.3 gives a design example.

The layout of the trench will depend on the shape of the land to be used and its slope as shown in Figure 10.3. If the site for the infiltration field is sloping, then laying pipes in parallel is not possible. The pipes should be laid in series across the slope of the land Figure 10.3. The sloping pipe connecting the infiltration trenches is usually raised at its inlet to prevent all the effluent running into the bottom trench and to ensure that the upper trenches are fully utilized before using the lower ones.

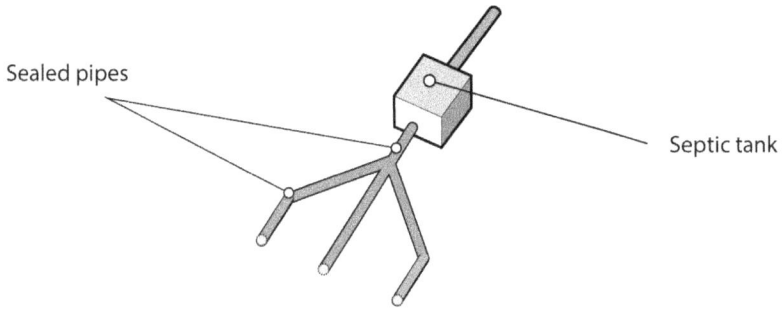

Infiltration trenches on a flat site

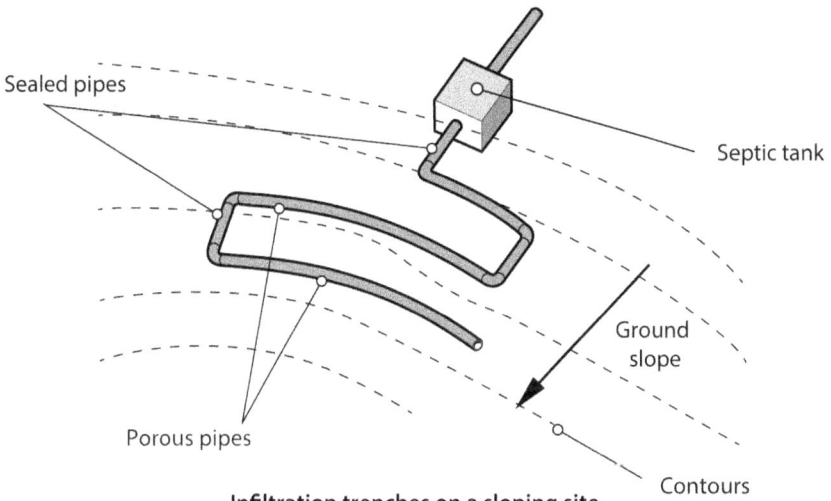

Infiltration trenches on a sloping site

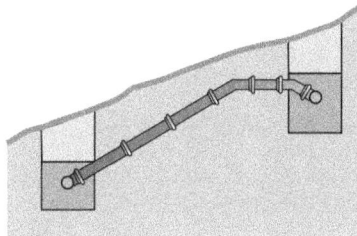

Connection between two infiltration trenches at different levels

Figure 10.3 Infiltration trenches

Box 10.2 How to make a pipe porous

There are a number of ways of making a pipe porous:

- The pipe walls can be made of a porous material such as concrete made with a reduced amount of sand.
- Commercially available porous pipes can be used, such as those used for land drains or drip-feed irrigation systems. There are also manufactured products that come complete with a gravel surround.
- Small holes or slots can be cut in the walls of the pipe (e.g. plastic land drains).
- The joints between pipes can be left unsealed (only for pipe lengths of around 1.0 m).

Table 10.1 Design data for infiltration trenches

Data	Value	Comments
Trench width	0.3–0.6 m	As narrow as possible, as infiltration is only through the side walls.
		Only use wider trenches if access to the trench during construction is required.
Trench depth	Approx. 1.0 m below distribution pipe invert	Depends on the infiltration area required, the space available, and the infiltration capacity of the soil.
		The trench bed should be above the water table as the infiltration process is less efficient underwater.
Distance between multiple trenches	1.0 m side wall to side wall minimum	
Pipe diameter	Depends on discharge volume but generally 100 mm	
Pipe slope	Horizontal	This will mean that the pipes nearest the inlet receive effluent first, but designing and accurately constructing pipes with a very shallow slope is very difficult
Bedding gravel	Around 4–10 mm diameter	Gravel should be washed to remove fine particles

---

Box 10.3 Calculating the length of the infiltration trench

The process is similar to that used for designing infiltration pits.

- Calculate the wall area required for infiltrating the wastewater (A)
- Decide the depth of trench below the effluent pipe (H)
- Calculate the total length of side wall required (C) = A/H
- The length of trench required is half of the total side wall length.

Note: Soil texture changes very quickly. Always measure the infiltration rate at a number of places within the drainage area.

(Continued)

Box 10.3 Continued

Case study
Design an infiltration trench to dispose of the effluent from the septic tank designed in Box 9.8.

Assume the soil is sandy loam and the water table is at a depth of 5 m but is highly saline and unsuitable for a drinking water supply.

Daily effluent flow rate = 2.0 m^3

Soil infiltration rate = 25 litres/m^2/day (Table 4.3)

Soil infiltration area = 2.0 × 1,000/25 = 80 m^2

Assuming that the infiltration trench is 1.0 m deep below the distribution pipe invert, then:

Trench length = 80/(1 × 2) = **40 m**

If this is judged too long, the trench could be deepened or the flow divided among multiple trenches.

Advantages
- No deep excavation.
- Makes use of soils near the surface that are generally more porous.
- Flexible layouts possible (pipes don't have to be laid in a straight line) to suit site conditions.
- Easier to maintain because everything is close to the surface.

Disadvantages
- Takes up a lot more space than a single infiltration pit.
- Careful design and construction essential to good operation.
- Frequent failures because of poor design and construction or failure of the previous treatment system.

10.1.3 Multiple trench systems

Subsurface drainage systems commonly fail. The main reasons are the drainage system is too small for the amount of wastewater to be infiltrated; or there is a failure to prevent the overflow of suspended solids into the distribution pipes (usually because of failure to maintain the previous treatment system), causing a blockage of the gravel bed and the soil interface.

One method of prolonging the life of the system is to construct two separate systems side by side (Figure 10.4). Each system is used, in turn, for a fixed period of, say, six months. This allows drainage fields to dry out between uses and the biological film to break away from the soil wall, unblocking the soil. Twin drainage fields are an expensive solution, requiring careful management. They are usually appropriate only for large installations. In smaller systems, the only options are either to construct another infiltration system nearby or to flush out the existing system. Whichever option is chosen, try to establish the cause of the failure first and put it right.

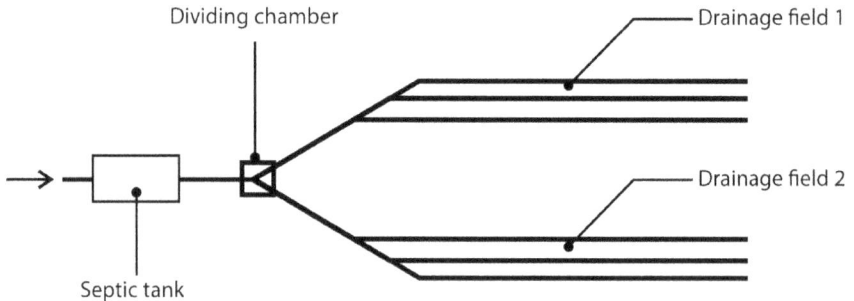

Figure 10.4 Multiple infiltration trench systems

10.1.4 Infiltration bed

Where the faecal sludge quantities are relatively small, it may be possible to separate the solids from the liquid and infiltrate the liquid into the ground at the same time. A drying bed can be constructed similar to the one shown in Figure 9.6 but without the impermeable base lining.

Since the soil infiltration process is aerobic, it is important that the bed is rested between episodes of infiltration to allow the liquor to fully disperse into the ground and air to return to the soil profile. The sludge drying time is usually sufficient for that to take place. Slurry application rates will generally be controlled by the subsurface infiltration rate rather than by the sludge drying parameters. All liquor should have been infiltrated within two to three days of application. Effluent infiltration quantities can be based on the figures given in Table 4.3 but, because of the likely organic solids build-up, it is suggested that the recommended figures are halved. Facilities for covering the beds should always be provided in case of rain. An example of the design process is shown in Box 10.4, it is essentially the same as that shown in Box 9.3.

Advantages
- Good dewatering efficiency, especially in hot dry climates.
- Simple technology using local materials and skills.
- Low capital cost (excluding land).
- Simple operation.
- Can partially treat excreta from storage vault-based toilets (such as chemical toilets, storage tank toilets, or septic tanks).

Disadvantages
- Requires a large land area.
- Unsuitable for large populations (see Box 10.4).
- Public health risks whenever excreta are handled, transported, or processed .
- Nuisance from odour and flies, especially where the influent is anaerobic (such as from a septic tank).
- Labour intensive, requiring constant supervision.

Box 10.4 Infiltration beds design for a population of 10,000

Assuming fresh faecal sludge is being collected in a low-income community where water is used for anal cleaning.

The daily sludge production rate is assumed as: 2.6 + 0.25 = 2.85 litres/person/day (Table 4.2).

Weekly production is approx. 20.0 litres/person/week

Daily total is 28.5 m³/day

Therefore, weekly total is 200 m³

Assume the subsoil is primarily sandy loam with an infiltration capacity of 25 litres/m²/day (Table 4.3)

Halve the infiltration capacity to allow for blocking = 12.5 litres/m²/day

Daily infiltration area = (28.5 × 1,000)/12.5 = **2,280 m²**

Assume a new bed is used every day (this can be changed if the bed sizes turn out to be too large or too small).

Assume, based on local experience, the drying time for sludge is three days.

Total drying cycle is loading time + drying time + emptying time

If it takes two days to empty, then:

Total drying time is six days, one of which is a non-working day.

The number of drying beds required is six, but add an extra bed to allow for maintenance and repair (Figure 10.5).

This is patently an unrealistic option for this volume of sludge (unless a huge amount of land is available) but it illustrates the design process.

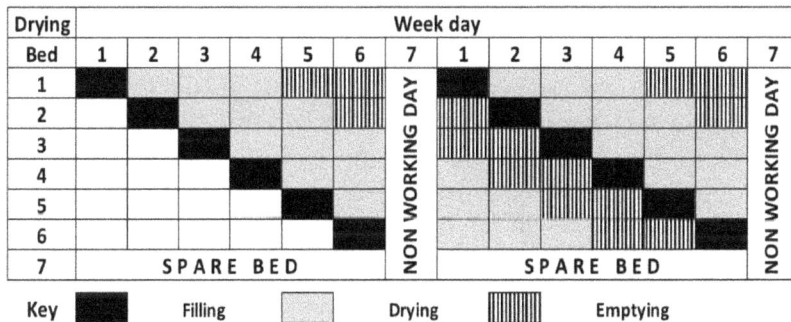

Figure 10.5 Drying bed usage
Source: Adapted from (** Tayler, 2018)

- Limited sludge stabilization and pathogen reduction.
- Unsuitable for a high water table, hard ground, or flooded environments.

10.1.5 Infiltration mounds

If the subsoil is impermeable (clay or rock) or the water table is close to the surface, the infiltration system can be placed in a mound, as shown in Figure 10.6. A mound of porous sand has, near its top, a gravel bed containing a network of porous distribution pipes connected via a pumping and dosing chamber to the effluent source, such as a septic tank. The original soil surface is disturbed (e.g. by ploughing) to break up its structure and improve

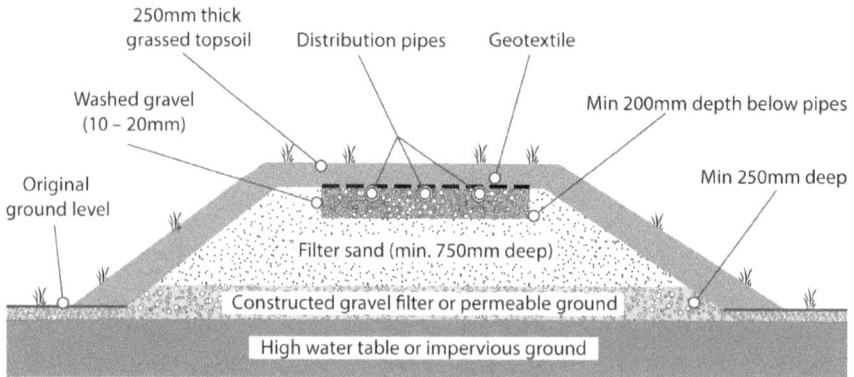

Figure 10.6 Cross-section through a mound infiltration unit
Source: Adapted from (Homeseptic Private Drainage Solutions, n.d.).

permeability. The distance to the water table or bedrock governs the height of the mound. The plan area of the mound is decided by the permeability of the natural topsoil. The area must be large enough to infiltrate the total volume of effluent produced.

Infiltration mounds are rarely suitable for the early stages of an emergency response because of the construction time and complex layout. However, if further information is required, refer to Appendix 7.

10.1.6 Sand filters

A single-pass sand filter system (Figure 10.7) is a secondary treatment after suspended solids have been removed. It filters the influent through sand before sending it for final treatment or direct discharge. Other similar filters use peat, pea gravel, crushed glass, or other experimental media, but sand is the best understood and the most predictable. Treatment mechanisms in a sand filter include physical filtering of solids, ion exchange (alteration of compounds by binding and releasing their components), and decomposition of organic waste by soil-dwelling bacteria. Although sand filters can be used as a final treatment process, they are most commonly used as an intermediate stage to reduce the organic loading on subsurface infiltration systems in low permeability soils or infiltration mounds. A properly operating sand filter receiving domestic-strength influent should produce high-quality effluent with less than 10 mg/litre BOD, less than 10 mg/litre total suspended solids (TSS), and less than 200 cfu/100 ml (Gustafson, et al., 2001).

Partially treated wastewater flows into a sump or lift tank. A pump introduces the influent at the top of the watertight sand filter, using pressure distribution to apply the wastewater evenly to the filter surface to maximize treatment. A timer is used to dose the entire surface of the filter intermittently with wastewater. This draws oxygen from the atmosphere through the sand

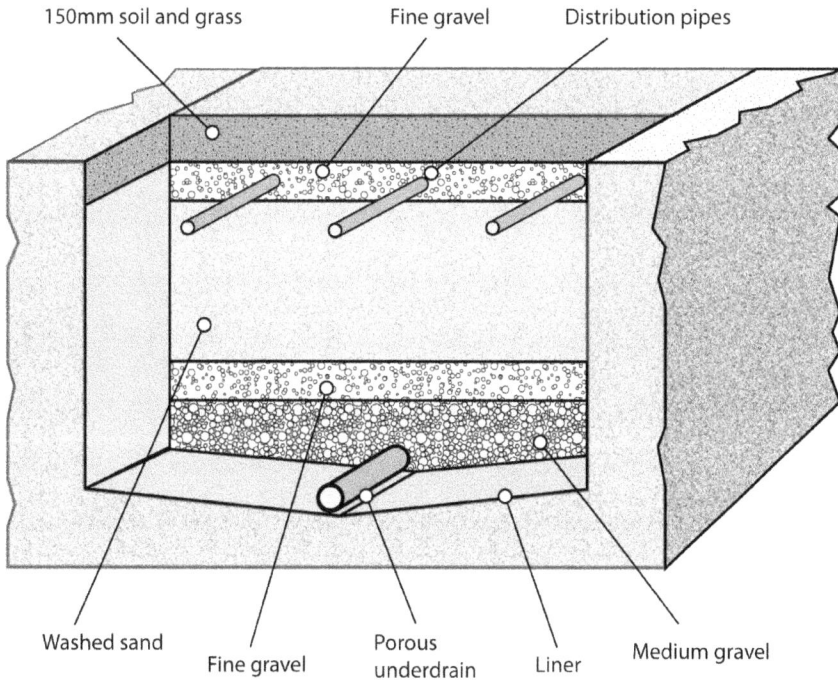

Figure 10.7 Section through a sand filter
Source: Adapted from (Gustafson, et al., 2001)

medium and its attached microbial community. Sand filter systems may also be successfully retrofitted into drainage fields that have failed because of excessive organic loading or from lack of maintenance. Sand filters are commonly used for treating effluent from small communities and institutions.

This treatment process is generally unsuitable for the early stages of an emergency because of the complexity and start-up period. However, further information is given in Appendix 7.

10.2   Urine as liquid fertilizer

Human urine has many uses from making medicines to formulating gunpowder. In emergencies, however, its most likely use is as a fertilizer. Stored urine coming from urine-diverting toilet systems is a concentrated source of nutrients that can be applied as a liquid fertilizer in agriculture (to replace chemical fertilizers) or as an additive to enrich compost. Urine contains most of the nutrients excreted by the body. Soluble substances in urine include essential plant nutrients such as the macronutrients nitrogen (N), phosphorus (P), and potassium (K), as well as smaller quantities of micronutrients such as boron (B), iron (Fe) and zinc (Zn). The nutrients in urine are in a form readily available to plants, similar to ammonia and urea-based fertilizers, and with

Table 10.2 Recommended storage time for urine based on estimated pathogen content and crops for larger systems

Storage temperature (°C)	Storage time (months)	Possible pathogens in the urine mixture after storage	Recommended crops
4	≤1	Viruses, protozoa	Food and fodder crops that are to be processed ^^^
4	≤6	Viruses	Food crops that are to be processed ^^
			Fodder crops ^
20	≤1	Viruses	Food crops that are to be processed ^^
20	≤6	Probably none	All crops^ ^^

Notes: ^Not grasslands for production of fodder.

  ^^ For food crops that are consumed raw, it is recommended that the urine be applied at least one month before harvesting and that it be incorporated into the ground if the edible parts grow above the soil surface.

Source: (WHO, 2013)

comparable results on plant growth. Table 10.2 shows the WHO guidelines for the storage of urine prior to use in agriculture.

Stored urine should not be applied directly to plants due its high pH. Instead, it can be applied directly to the soil before planting, by pouring into furrows or holes at a sufficient distance away from plant roots and then immediately covered. Alternatively, it can be diluted (1:10–1:15) and used frequently on plants as a general fertilizer. The advantages of dilution are a noticeable odour reduction and a decreased risk of over-application. Urine application is not considered a priority in acute emergencies, but it might be an option during the stabilization and recovery phases.

Advantages
- Excellent natural fertilizer.
- Minimal risk of infection, especially when stored for a month.
- Especially useful for improving compost made with human faeces.

Disadvantages
- Strong odour from stored urine.
- High pH makes it unsuitable for direct application to plants.
- Long storage time prior to application.

Section adapted from Robinson, 2010.

10.3   Evaporation

Many of the treatment processes already described contain an element of liquid reduction through evaporation, such as drying beds, planted drying beds, and sedimentation and thickening ponds. Mechanized processes for evaporating partially treated effluent are rarely feasible in emergencies: they are energy dense, suffer from blockages from the suspended solids, and may disperse volatile compounds containing pathogens into the atmosphere, causing a health hazard. However, in hot, dry, and windy climates, natural evaporation may be appropriate as a short-term emergency solution, especially where the moisture deficit (evaporation minus rainfall) exceeds 750 mm annually. Evaporation ponds are suitable only for effluent with low suspended solids as sludge accumulation will restrict evaporation.

Ponds are usually sized to provide the necessary surface area to evaporate all the effluent produced by earlier treatment processes plus the rainfall that falls on the pond. However, in emergencies, ponds are likely to be used only for short periods, so such calculations are unnecessary. However, Box 10.6. outlines the design of such a pond. Emergency ponds can take advantage of temporary climatic conditions such as summer heat and/or periods of low rainfall and high winds.

Land must be naturally flat or be shaped to provide ponds with uniform depth. The liquid depth is commonly 1–2 m, but it could be shallower if space allows. The surface shape can be any that fits the available land. The total surface area is a function of the average daily inflow and the evaporation rate that can often be obtained from local or national hydrological and meteorological organizations. Failing that, it can be approximated by carrying out a pan test (Box 10.5). Evaporation ponds are similar to sedimentation and thickening ponds (Figure 9.5) but without the outlet pipe and with basin

---

Box 10.5 Pan test to estimate evaporation rates

A shallow dish is filled with water and the daily drop in water level measured to estimate evaporation. Different institutions have developed different evaporation tests as the shape, size, and location of the dish have a significant impact on the overall result. The most widely used is that produced by the United States National Weather Service: the Class A evaporation pan.

This is an unpainted galvanized iron cylinder with a diameter of 1,207 mm and a depth of 255 mm. The pan rests on a carefully levelled, wooden base 150 mm above ground (Figure 10.8). The pan must be covered with an open mesh metal frame to keep out animals. Evaporation is measured daily as the depth of water (in millimetres) evaporates from the pan. The measurement day begins with the pan filled to 50 mm from the pan top. At the end of 24 hours, the amount of water to refill the pan to exactly 50 mm from its top is measured (Mishra, n.d).

If rainfall occurs in the 24-hour period, it can be taken into account in calculating the evaporation; in practice, it is simpler to ignore that day's reading. Evaporation cannot be measured when the pan's water surface is frozen. Readings are taken over a number of days and the evaporation rate averaged.

In practice, measuring evaporation losses accurately in an emergency is difficult. Allowing the water level to drop for two to three days is less accurate but is easier to measure.

The evaporation area required = C_p × [effluent flow rate (litres/day)/evaporation rate (mm/day)]

C_p is the pan coefficient. In the absence of more accurate data, this is usually taken as 0.7.

(Continued)

Box 10.5 Continued

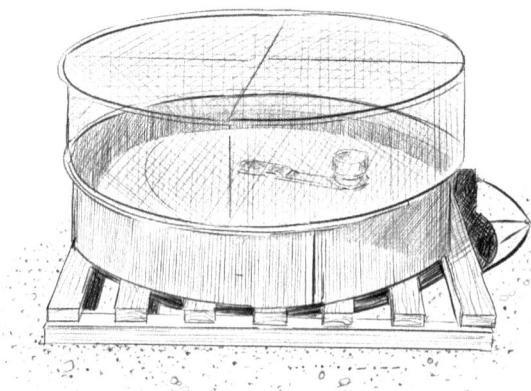

Figure 10.8 Simple evaporation pan
Source: Adapted from (Bidgee, n.d.)

Box 10.6 Evaporation pan case study

Calculate the size of the evaporation basin required to dispose of the final effluent from the septic tank designed in Box 9.8. Assume that a locally made evaporation pan measures a drop in water level in the pan of 15 mm over three days. However, rain can be expected on average one day per week, with a daily rainfall of approximately 4 mm per day.

Average weekly drop in water level in the evaporation pan on dry days is 15/3 = 5 mm

From Box 9.8, the average daily flow is 2 m³ = 14 m³/week

Total volume of liquid entering the pan a week (V) = weekly effluent flow plus the rainfall = 14 + (A × 0.004) m³

Where 'A' is the area of the evaporation pond (m²)

Daily evaporation volume (one rainy day a week) = V/6

Minimum surface area required (A) = Daily evaporation volume ÷ Daily evaporation rate
= (V/6)/0.005
= (14 + 0.004 A)/(6 × 0.005)
0.03 A = (14 + 0.004 A)
A = 538 m²

The depth of the tank is not important but 0.3–0.5 m is suggested as it is difficult to construct a perfectly flat bed of such a size. It will also allow for variations in evaporation and rainfall rates.

In practice, it is probably wise to construct the tank 25–50 % larger to allow for variations in flow and evaporation rates, higher frequencies of rainfall, and the likelihood that actual flows will be higher than designed for.

walls as low as possible to maximize any wind effect. They should be located away from habitation because of odour and insect issues and must always be fenced to keep out animals and humans.

Advantages
- Simple and quick to set up.
- Low maintenance.

- Suitable for all ground conditions provided bed is lined with permeable soils.
- No groundwater contamination (if the pan bed is sealed).

Disadvantages
- Strong odours.
- Often requires a large flat area.
- Subject to the vagaries of the weather.
- Potential health hazard, especially to children if not properly fenced off.

10.4   Chlorination

Chlorination of wastewater effluent with low levels of suspended solids is widely used to ensure the destruction of pathogens prior to discharge. Disinfection rates vary depending on the effluent but are usually in the range of 5–20 mg/litre. The contact time also varies, but it is recommended to use the same as for water supply disinfection (30 minutes minimum) until laboratory tests can be undertaken. Once disinfection is complete, any remaining chlorine must be removed prior to discharge to prevent environmental damage in the receiving waters. Chemicals such as sulphur dioxide and sodium bisulphite are commonly used for this purpose. See (United States Environmental Protection Agency (EPA), 1999) for further details.

10.5   Disposal to surface waters

10.5.1 Sewer outfalls

It is frequently necessary to discharge sewage effluent to surface water such as a river, lake, or the sea. When installing such an outfall, consideration must be given to its impact on the receiving waters. The main impacts to be considered are:

- increased bed erosion leading to an increase in sediment build-up downstream;
- accumulation of sediment immediately upstream of the outfall;
- trapping of fish because of the turbulence created by the outfall;
- direct loss of bankside and bed habitat for plants and animals.

In addition to these engineering issues, there are issues relating to the quality of the sewage effluent, which, if not properly treated, can cause extensive and long-term environmental damage.

Discharging contaminated water into surface water is extremely dangerous to the health of those living nearby and downstream, and to the environment. It should be sanctioned only in the most extreme of circumstances.

Most countries have their own standards for effluent discharge quality and it is the duty of the implementing agency to be aware of them and attempt to meet them. Where this is not possible, written approval for not doing so should be obtained.

Design

There are a number of different designs of sewer outfall, as described in Table 10.3. Key features to consider when locating them are:

- a stable bank;
- no evidence of erosion, undercutting, or sediment deposition in the water course;
- a straight section of river;
- minimum impact on wildlife habitats;
- in stationary waters, the pipe soffit (top of the sewer pipe) is below the minimum water level.

Table 10.3 Sewer outfalls

Outfall type	Suitability	Key features
Submerged outfall	Suitable for discharges that require immediate dilution or for high-volume/velocity discharges. They have a low hydraulic impact on the receiving waters.	• Soffit of the outfall should be below low water level • Outfall pipe should not protrude beyond the bank line • Pipe should be buried beneath the bank and native vegetation reinstated • A sampling chamber should be incorporated to check effluent quality
Partially submerged outfall	The outfall can be fully submerged during periods of high flows in the receiving waters but is generally only partially submerged. They are suitable for high-velocity discharges and have a low hydraulic impact on the receiving waters.	• Outfall pipe should not protrude beyond the bank line • Pipe should be buried beneath the bank and native bank vegetation should be re-established • Safe access should be provided • Effluent sampling chamber should be provided • Mitred headwall is flush with the bank, allowing for easy maintenance and reducing trip hazard
Bankside outfall	Only recommended where there is inadequate water depth for submerged outfalls or the outfall is set back from the edge of the water course.	

(Continued)

Table 10.3 Continued

Outfall type	Suitability	Key features
	These can have a high hydraulic impact on the receiving water.	• No part of the outfall structure should protrude beyond the line of the bank – this includes headwalls, wing walls, and protection aprons
		• Should have a silt apron to aid silt removal or a raised inlet to avoid silt build-up in the pipe
		• Safe access should be provided
		• Sample chamber should be provided (if necessary)
		• Native bank vegetation should be re-established after construction
Setback outfall	Suitable for low-velocity clean water discharges. The outlet pipe should finish some distance from the entry to the watercourse to provide a surface discharge route and semi-natural entry (e.g. it should discharge to a setback channel or wetland). These have a low hydraulic impact on the receiving waters.	• Outfall is setback from the bankside and water edge • Erosion protection should be minimal as low velocities are involved • Bankside and wetland can be planted with local native species, thus reducing the impact and maximizing biodiversity • The constructed channel bank profile should have a maximum slope of 1:3 for management and health and safety considerations • Safe access should be provided

Correct alignment and design of the outfall can help reduce scour around the structure and erosion of the bed and banks. In rivers, the discharge should be in line with the flow, as this helps reduce turbulence and erosion. No part of the outfall structure should protrude beyond the line of the bank; this includes headwalls, wing walls, and protection aprons. The structures should be located above peak river height with an apron on the bank below the outfall to protect it from erosion (Figure 10.9).

Advantages
- Recognized alternative to disposing of faecal waste effluent where ground infiltration is not possible.
- Design and construction of discharge points are widely understood.
- Generally simple and quick to set up.

Disadvantages
- Contaminated effluent is a danger to health and the environment.
- Obtaining formal approval can be slow and bureaucratic.
- Emergency outfalls frequently damaged by unexpected flash flooding.

Section adapted from Scottish Environment Protection Agency, 2019.

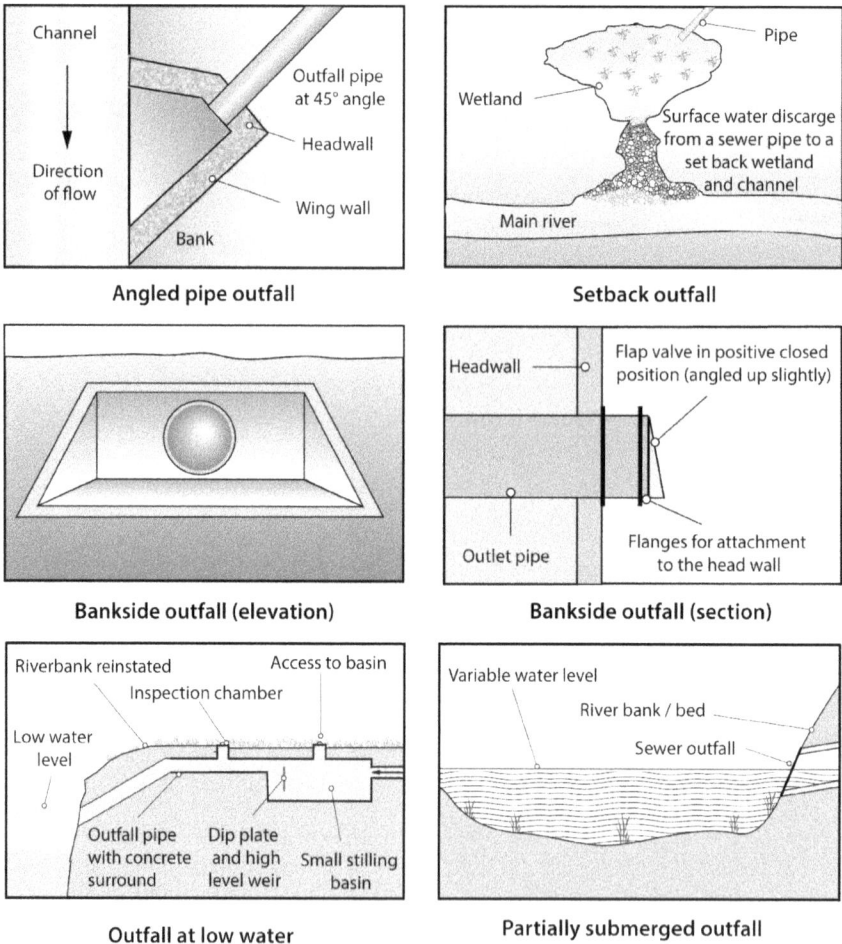

Angled pipe outfall

Setback outfall

Bankside outfall (elevation)

Bankside outfall (section)

Outfall at low water

Partially submerged outfall

Figure 10.9 Effluent discharge into a river
Source: Adapted from (Scottish Environment Protection Agency, 2019)

10.6 Disposal of treated faecal sludge

The WHO provides clear guidelines on the safe use of faecal sludge in agriculture. Full details can be found in (World Health Organization, 2006). The key guidance is summarized in Table 10.4.

10.6.1 Dehydrating faeces

Fresh faeces captured below a toilet in a porous container and stored in the absence of moisture will gradually dehydrate. Dehydration is very different from composting as the organic material is not degraded or transformed; only the moisture is removed. This can reduce the volume of faeces by about 75 per cent.

Table 10.4 Recommendations for storage treatment of dry excreta and faecal sludge before use at the household and municipal levels

Treatment	Criteria	Comment
Storage; normally 2–20°C	1.5–2 years' storage	• Eliminates bacterial pathogens
		• Regrowth of E. coli and salmonella may need to be considered if rewetted
		• Will reduce viruses and parasitic protozoa below risk levels
		• Some soil-borne ova may persist in low numbers
Storage; normally 20–30°C	Greater than 1 year's storage	• Almost total inactivation of viruses, bacteria, and protozoa
		• Inactivation of Schistosoma eggs (greater than 1 month)
		• Inactivation of nematode eggs, e.g. roundworm, hookworm, whipworm
		• Survival of certain percentage (10–30 per cent) of Ascaris eggs (greater than 4 months)
		• More or less complete inactivation of Ascaris eggs will occur within 1 year
Alkaline treatment	pH greater than 9 for more than 6 months	• If the temperature is greater than 35°C and moisture less than 25 per cent, lower pH and/or wetter material will prolong the time for absolute elimination

Source: Adapted from (WHO, 2013).

Completely dry faeces are a crumbly, powdery substance. The material is rich in carbon and nutrients, but can still contain worm eggs, protozoan cysts, or oocysts (spores that can survive extreme environmental conditions and be reanimated under favourable conditions) and other pathogens. The degree of pathogen inactivation will depend on the temperature, the pH, and the storage time. It is generally recommended that untreated faeces should be stored and dehydrated for between 18 and 24 months where the average temperature is between 2 and 20°C or for 12 months at average temperatures above 20°C, although pathogens can remain viable even after this time. The retention time can be reduced to around six months if the pH is raised above 9. See (WHO, 2013) for more specific guidance. The dehydrated faeces can be used as an additive in composting, mixed directly into the soil, or buried elsewhere if reuse is not intended. Extended storage is also an option if there is no immediate use for the material.

10.6.2 Pit toilet sludge/humus

Pit toilet sludge experiences almost no increase in temperature because the conditions in the pit (the presence of oxygen, moisture, and the carbon to nitrogen ratio) are not optimized for the composting processes. It is therefore referred to as pit humus rather than compost. The texture and quality of pit humus depend on the materials that have been deposited in the pit in addition to excreta and storage conditions. It may contain non-organic materials such as plastic, cloth, metal, etc. that must be removed prior to reuse. Pit humus contains fresh excreta that may include pathogens. The sludge should undergo treatment or prolonged storage as recommended in Table 10.4.

Pit humus adds nutrients and organic content to the soil and improves the soil's ability to store air and water. It can be mixed into the soil before crops are planted, used to start seedlings or indoor plants, used to plant trees, or simply mixed into an existing compost pile for further treatment. It has been shown to increase the availability of micronutrients and to contribute to the overall food security, resilience, and well-being of the affected community as part of community greening programmes when mixed with the soil used for food production. Where food production is not an option, pit humus can be used to restore land where natural events have removed the top layer of the soil. Utilizing pit humus is appropriate for the stabilization and recovery phases of an emergency (** Gensch, et al., 2018).

10.6.3 Bulk faecal sludge disposal

Post-treatment faecal sludge comes in a variety of formats depending on the previous treatment processes. It can include liquefied sludge from ABR plants, sewage treatment plants, and septic tanks, partially dried sludge from pit toilets, and dry sludge from evaporation and drying beds. No matter what the post-treatment process is, it cannot be assumed that the sludge is completely free of pathogens and therefore its final disposal should not produce a health hazard.

Dry sludge cake can be buried to provide a nutrient source for shrubs and trees. The sludge should be buried either below or adjacent to the root ball (Figure 10.10). The plant can then grow its roots into sludge to extract nutrients as required. Sludge can also be buried in this way without a link to planting. Check for potential groundwater pollution prior to burial.

Where liquid sludge has undergone advanced treatment, such as post-treatment storage (Table 10.4), doesn't contain harmful contaminants such as heavy metals, and has a low pathogen load, it can be used as a surface dressing on farmland. Where possible, the sludge should be ploughed into the ground to further reduce potential health hazards.

10.6.4 Co-composting

Co-composting is the controlled aerobic breakdown of organic material, using more than one waste source. Faecal sludge has a high moisture and nitrogen

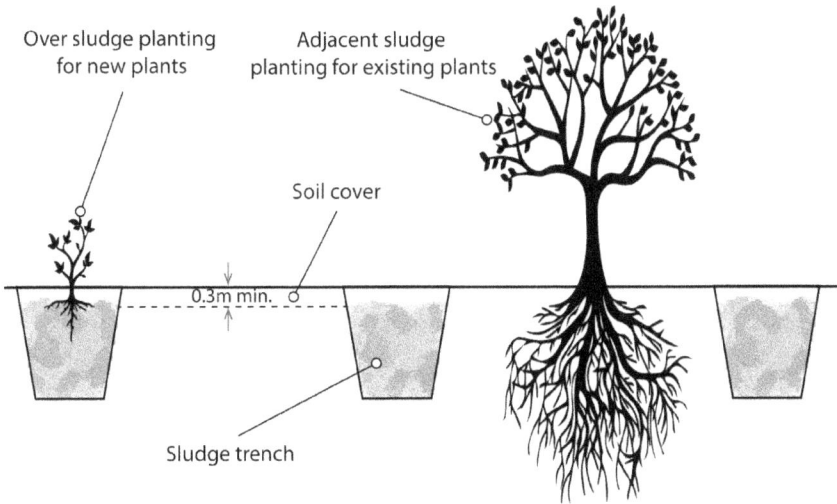

Figure 10.10 Over and adjacent planting to sludge filled trenches

content, while organic solid waste (from food or agricultural waste) is high in organic carbon and has good bulking properties, which promote aeration. By combining the two, the benefits of each can be used to optimize process and product. The mix is composted in such a way as to raise the temperature to greater than 55–60°C. This requires the careful control of moisture content, the carbon to nitrogen ratio, and aeration. The resultant mass results in the elimination of pathogens and rapid decomposition of the waste material. The process produces a safe, stable end product that can be used as a compost or soil amendment. Three commonly used methods of co-composting are:

- **open-windrow**: the mixed waste is piled into long heaps called windrows and left to decompose (Figure 10.11);
- **in-vessel**: controlled moisture, air supply, and mechanical mixing;
- **a combination of open-windrow and passively aerated static pile**: waste sits in a static pile for around two to three months and then is moved to windrows for further decomposition.

Key components in the design of a co-composting facility include space for sorting and waste separation, drying beds, composting units, screening, storage of compost and discards, hygiene and disinfection infrastructure, on-site wastewater treatment system, staff facilities, and a buffer zone. Depending on the climate and available space, the facility may need to be covered. The facility should be located close to the sources of organic waste and faecal sludge to minimize transport costs, but still a distance away from living areas to minimize any perceived or real health risks. Windrow piles should be at least 1 m high and insulated with a 30 cm layer of compost, soil, or grass cuttings to promote an even distribution of heat. In colder climates, heaps work best at 2.5 m high and 5 m wide. Windrows often

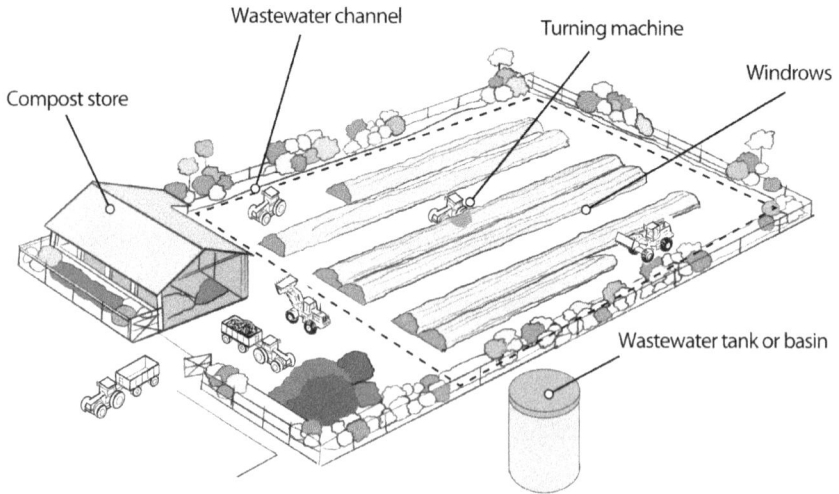

Figure 10.11 Co-composting plant using windrows
Source: Adapted from (Barnard Health Care, 2022)

include lengths of perforated piping to improve ventilation in the centre of the pile (Figure 10.12). Sludge must have a low moisture content such as that taken from a urine diversion toilet or drying beds prior to mixing with organic waste. Composting takes place on an impermeable surface (such as concrete) to collect the leachate, which can then be reintegrated into the piles or treated separately. The composting process needs to be carefully managed to ensure that pathogens are being destroyed. Thermophilic composting for one month

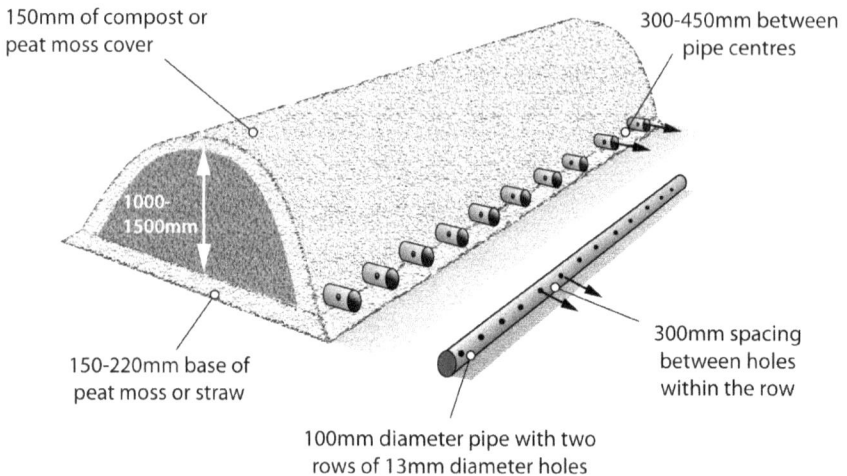

Figure 10.12 Aerated windrow
Source: Adapted from (Rynk, 1992)

at 55–60°C is recommended, followed by two to four months of curing to stabilize the compost.

Because of the high level of organization and labour needed to sort organic waste, manage the facility, and monitor treatment efficiency, this technology is unlikely to be practical in the acute response phase. However, it can be considered a viable option in the stabilization and recovery phases of an emergency. Experience has shown that co-composting is a viable method for the treatment of fresh faeces provided the urine is collected separately. Facilities operate best when they are established as a business with compost as a marketable product that can generate revenue to support cost recovery. However, compost sales cannot be expected to cover the full cost of the service

The wastes generated by a single family are not enough to support such a large increase in temperature and therefore the process cannot be relied on to destroy pathogenic organisms. Any toilets based on this process must contain other features to prevent the spread of disease, the most common one being the storage of the organic mass for at least one year (see Table 10.4) to ensure the destruction of pathogens. This can be achieved in twin pit toilets (see Chapter 6, section 2.5) and double-vault urine diversion toilets (Chapter 6, section 2.8). However, household composting toilets are not recommended for the acute response phase of emergencies: they are expensive, difficult to operate, and largely unacceptable to users who have not used them before. They may be an option for the stabilization and recovery phases. Further information on composting is given in Chapter 15, section 11.3.

10.6.5 Landfill disposal

Landfill disposal of faecal sludge involves collecting the contents of household, communal, or public toilets and dumping it with refuse at an existing landfill site. Generally, landfill disposal of faecal sludge should be considered only during extreme emergency conditions as a short-term solution where no alternative method of disposal is possible. It is considered very bad practice to mix waste streams (municipal wastes and human wastes) and a much better solution is to identify a suitable site for a dedicated faecal sludge treatment process.

A landfill site is an engineered pit, in which layers of municipal waste and sometimes faecal sludge are deposited, compacted, and covered with topsoil. The site has an impermeable lining (e.g. puddled clay or butyl sheeting) on the base to prevent groundwater pollution, a leachate collection and treatment system, groundwater monitoring, gas extraction (the gas is flared or used for energy production), and a cap system. The capacity is planned and the site is chosen based on an environmental risk assessment study (United Nations Environment Programme (UNEP), 2002). Landfills need expert design as well as skilled operators and proper management to guarantee their functionality. It is very unlikely that such a site would be specifically constructed during the acute phase of an emergency, but existing sites could be inspected.

Ideally, at the end of each day, about 0.5 m of compacted soil should be placed over the waste. This helps to prevent animals and vermin from disturbing the waste, prevent flies from breeding, minimize odour, and prevent wind-blown waste (e.g. plastic bags or other articles that may contain pathogens). It also helps to exclude rainwater and thus minimize leachate.

Data does not exist specifying the amount of faecal sludge that can be added to domestic garbage. However, the most likely controlling factor is the overall moisture content of the combined mass. The moisture content of the landfill should be in the range of 35–45 per cent water by weight. A lower moisture content will restrict biological decomposition and a higher level will lead to excess leachate. Further details can be found in Chapter 15, section 11.1.

Advantages

- Effective temporary disposal method for faecal sludge if well managed.
- Fast degradation of organic matter if correctly designed.

Disadvantages

- Fills up quickly if waste is not reduced and reusable waste is not collected separately and recycled.
- Requires a reasonably large area.
- Risk of groundwater contamination if not sealed correctly or if the liner system is damaged.
- High cost for high-tech landfills.

10.6.6 Incineration

The incineration of dried human faeces should be considered only where there is no alternative for its disposal. Unless conducted correctly, it has the potential to disburse pathogens over a wide area as well as creating nuisance from odour and smoke. Dried sludge from general faecal sludge such as pit toilets or septic tanks is not suitable for incineration because of the low percentage of organic matter, non-faecal waste, and high silt levels. The dried faeces should contain at least 60 per cent organic matter and have a moisture content below 30 per cent.

In the acute phase, a temporary incinerator such as that shown in Figure 10.13 could be considered. The drum can be fitted with a chimney, which should be at least 1 m taller than the surrounding structures. This helps to remove smoke and reduce the effects of pollution. The incinerator can be built directly above a sealed pit so that the ash can be emptied from the base of the drum and deposited directly into the pit below. Alternatively, the ash can be removed and buried nearby. Strictly speaking, this is not an incinerator

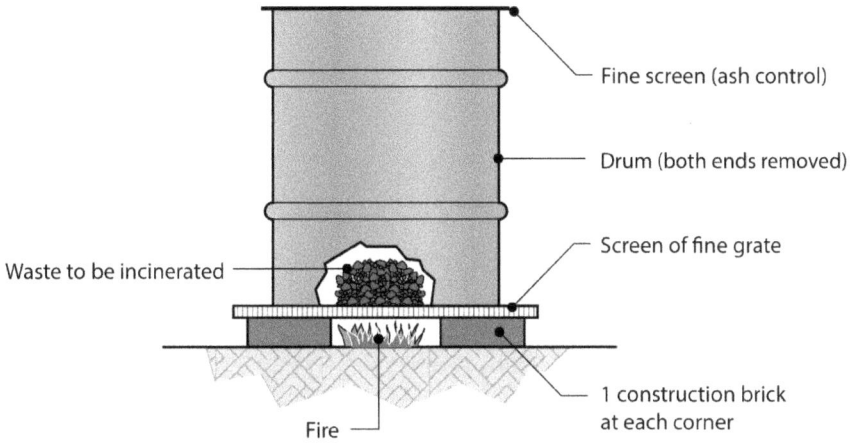

Fine screen (ash control)

Drum (both ends removed)

Screen of fine grate

Waste to be incinerated

1 construction brick at each corner

Fire

Figure 10.13 Temporary drum incinerator

but a burner, since it is unlikely to reach temperatures necessary to reduce all waste to ashes.

Basic incinerators such as this often cause serious problems with emissions and should always be positioned at least 100 m from all habitable buildings and food stores to minimize the effects of smoke pollution (** Harvey, et al., 2002).

For a longer-term solution, a properly designed incinerator such as that shown in Figure 10.14 can be constructed. This is designed to operate at temperatures of 1,000°C and above and will reduce even metallic waste to a fine uniform ash. The one shown was designed for the incineration of medical waste, but it is probably appropriate for faecal sludge with the addition of extra combustible material. Complete details of the design and construction of this incinerator can be found at (Picken, 2004).

Advantages
- In a properly designed incinerator, pathogens are totally destroyed.
- Sludge waste is totally mineralized.
- Heat is produced that can be harnessed for other purposes.
- Minimal odour.
- Low maintenance cost.

Disadvantages
- Temporary incinerators can have serious emission problems (smoke, odour, and airborne pathogens).
- Incineration incomplete in temporary incinerators and in efficient incinerators operated poorly.
- Requires skilled operators to work efficiently.

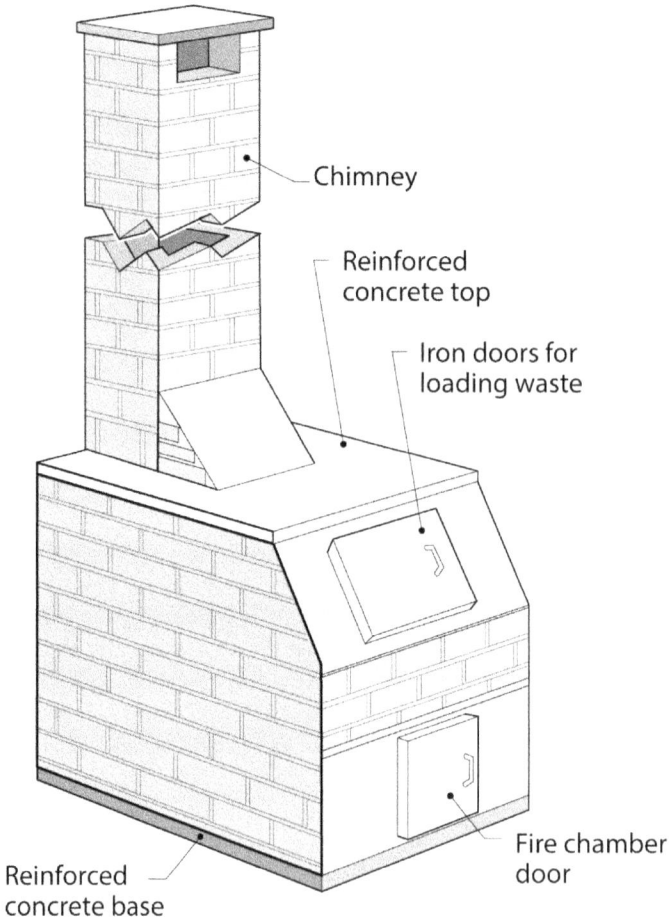

Figure 10.14 De Montfort medical waste incinerator
Source: Adapted from (Picken, 2004)

10.6.7 Pyrolysis/biochar

Pyrolysis is the heating of organic materials (including faecal sludge) at high temperatures in the absence of oxygen. The process may be classified as fast, intermediate, or slow, depending on the temperature used. Only the slow process is discussed here. This requires the organic material to be retained for a number of hours at a minimum temperature of 200°C but typically up to around 700°C. Pyrolysis differs from combustion (e.g. incineration) in that little or no carbon dioxide is released during the process. Organic material instead undergoes conversion into carbon in the form of hard, porous charcoal. This material, which is called biochar, can be used as a soil amendment or as a fuel source. The pyrolysis process produces

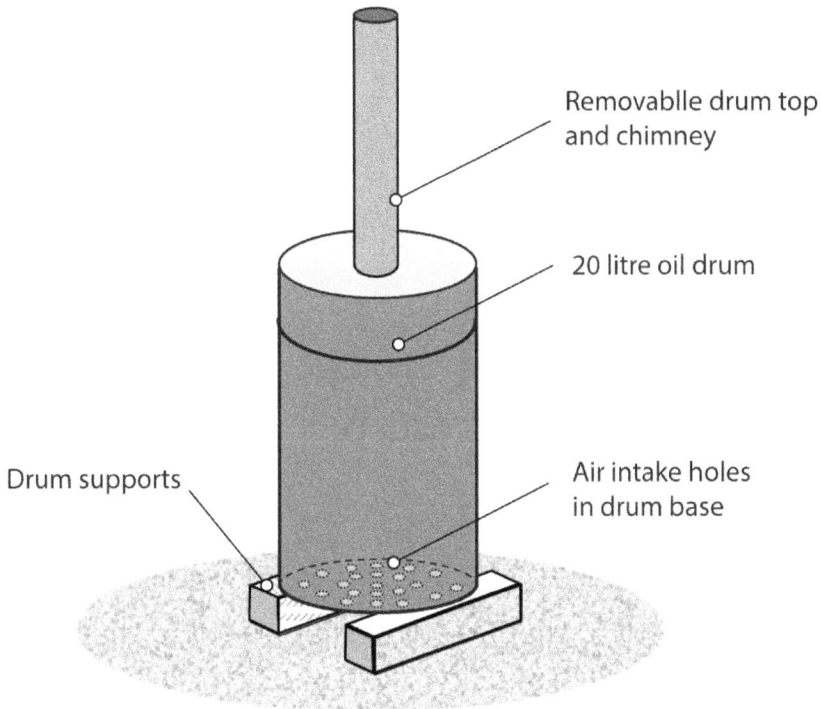

Removablle drum top and chimney

20 litre oil drum

Drum supports

Air intake holes in drum base

Figure 10.15 Simple biochar reactor

a mixture of gases that can be captured and used as the fuel to power the process. The process destroys all pathogenic organisms and increases the soil's ability to retain water and release nutrients more slowly. Biochar has been shown to lead to an increase of 25 per cent in crop yield at an application rate of 15 tonnes per hectare in tropical conditions, particularly in low nutrient acidic soils, but a small decrease in yield in temperate latitudes (** Tayler, 2018).

Design
The faecal sludge should have a solids content of around 60–70 per cent if it is to be self-sufficient in energy. In practice, a higher percentage is an advantage. Most small-scale pyrolysis plants operating in low-income countries operate in batch mode (Figure 10.15). This simplifies the technology and operational requirements but increases the need for an external fuel source to heat the reactor contents to the required reaction temperature.

Several pilot-scale initiatives have focused on the possible use of faecal sludge biochar to produce solid fuel briquettes. Box 10.7 provides brief information on one of these. Although there are many examples of biochar production on the internet, many with detailed design and construction advice, it is likely

---

Box 10.7 Biochar production from faecal sludge using pyrolysis

To date, most initiatives using pyrolysis to produce biochar or fuel briquettes from faecal sludge have been on a pilot scale. One such initiative is operated by Water for People with support from the Water Research Commission (WRC) in Uganda and involves production of sludge briquettes. Prior to pyrolysis, the incoming faecal sludge is dewatered on unplanted drying beds to a solids content of approximately 60 per cent and then further dried on racks to achieve a solids content of 80 per cent, which is suitable for the pyrolysis process. Currently, the organization is experimenting with two types of small kilns that have previously been used for carbonization of wood: a masonry-insulated retort kiln and a metallic kiln. The process involves the following steps:

1. Start-up fuel (wood or charcoal) is burned at the base of the kiln.
2. Dried sludge is added until the kiln is full.
3. Additional sludge is added as the sludge burns down (four to five hours).
4. When the fire penetrates the topmost sludge, the unit is air-locked to allow the pyrolysis process to continue overnight.

After the final step of the process, the carbonized biochar is crushed into fine particles and then blended with a binder such as cassava or molasses. Clay may also be added as a filler to reduce the burning rate of the briquettes, although this may not be necessary as the lack of pit lining means that sludge may already contain a high proportion of filler. Crushed charcoal can be added to increase the energy content of the mixture. After blending and adding water to increase the moisture content, the briquettes are produced using a mechanized extruder, screw extruder, hand/manual press, or honeycomb press.

    The calorific value of the briquettes is reported to be 7.5–15.5 MJ/kg compared with a calorific value of 12.5 MJ/kg for charcoal dust. The organization reports the selling price of charcoal to be 5.8 times the selling price of the briquettes, although it is not clear what the revenue and operating costs are for the system.

Source: (** Tayler, 2018)

---

that the focus of initiatives involving pyrolysis of faecal sludge will be on pilot-scale projects that are designed to explore the technical and financial viability of the option. Clearly, the latter will depend on demand for biochar and the existence of effective marketing systems.

Advantages
- Produces a pathogen-free product.
- End product suitable for burial or agricultural spreading.
- Provides an excellent and safe soil conditioner with sometimes high fertilizer value.
- On a large scale it is a virtually external energy-free process.
- Reduces the volume of sludge to be disposed.
- Slow carbon breakdown may allow the cost of production to be recovered from carbon sequestration credits.
- Potentially a suitable option for manually collected pit latrine sludge because of its high solids content .
- Pilot-scale plants are easy to fabricate from local materials.

Disadvantages
- Management difficulties controlling the quantity and quality of liquors and fumes.
- Faecal sludge requires pre-treatment.
- No concise design criteria for carbonizing faecal sludge, especially on a large scale.
- Probably unsuitable for very humid climates because of the difficulty of drying faecal sludge.
- Unsuitable for the acute response phase but possibly a consideration for later response phases.

10.7   Decommissioning

In some scenarios involving temporary facilities or as part of an exit strategy (see Chapter 3, section 7), it may be necessary to develop a programme for dismantling and decommissioning emergency toilets and faecal sludge management facilities. By default, the organization responsible for infrastructure construction is also responsible for decommissioning unless that role has been formally transferred to another agency. Some key issues to consider in decommissioning are outlined in Box 10.8

Box 10.8 Issues to consider when decommissioning

- Decommissioning should ideally be carried out during the 'dry' season when toilet pit or tank contents will have had the most opportunity to dry out. It is also the best time to decommission faecal sludge management facilities.
- Staff should be trained and provided with protective clothing for dismantling cubicles and removing toilet floor slabs and other toilet infrastructure. Hydrated lime or another form of disinfectant should be used to clean all materials dismantled to mitigate against unpleasant odours and health hazards from reuse. Cleaning should take place immediately after demolition or the wastes stored securely, as they are very vulnerable to theft by the affected population as construction material.
- Toilet pits and trenches should be backfilled with excavated soil. Lime can be added to the pit contents prior to backfilling to aid decomposition, although this is not normally necessary. If the pit contents are liquid, it may be necessary to dig an overflow trench from the top of the pit or tank to absorb displaced fluids. In most cases, the displaced liquid will gradually infiltrate into the surrounding soil or evaporate. Failing that, it should be removed for treatment prior to disposal. The pit or tank should then be capped with a mound of soil and rubble to allow for further settling of contents and to prevent excavation by animals and vermin. Capping the pit with a concrete slab should be considered in populated areas where interference is possible.
- Cement and other non-recyclable debris from the toilet or other dismantled facilities can be thrown into the pit prior to backfilling or taken away for off-site disposal in line with local authority regulations. Some materials such as plastic and wood may be recyclable, but they must be fully cleaned and disinfected before leaving the site.
- Vegetation such as grasses and trees can be planted on decommissioned sites where appropriate for site rehabilitation.
- If possible, the area should be fenced off to prevent it from being disturbed.

(Continued)

Box 10.8 Continued

- Sludge transport vehicles should be thoroughly cleaned (with steam or pressure wash cleaning, for example) prior to being handed over, put into storage, or moved to another site for reuse.
- All equipment and materials in sludge transfer stations should be cleaned and removed from the site. The site should be returned to its original condition with any contaminated land treated or removed.
- Temporary treatment plants must also be decommissioned. Partially treated faecal sludge and liquor must be moved to another treatment plant, earthworks levelled, equipment cleaned and removed, and imported materials such as filter material either cleaned prior to removal or buried. The site should be returned to its original condition unless given alternative orders.
- All staff personal protective equipment (PPE) should be properly laundered and returned to the staff or incinerated to prevent cross-contamination.

CHAPTER 11
Effects of low temperatures on toilets and faecal sludge management

Cold climates add a host of additional complications, including freezing of water and excreta, problems related to excavating frozen ground, and slowing down of biological processes. There is no agreed definition of what constitutes a cold region but for the purposes of this book it is defined as a region where the mean monthly temperature is below 1°C for one month or more per year. There are places where the soil freezes and thaws seasonally on every continent, including more than 50 per cent of the northern hemisphere's land surface. No matter how a cold region is defined, faecal sludge management (FSM) for large numbers of people is affected by cold temperatures. In 2015, more than 2.5 billion people lived in countries where the average monthly temperature has historically been below 1°C for at least one month of the year. Of these, 1.5 billion lived in countries where the average monthly temperature was below –5°C for at least three months of the year. At such cold temperatures, it is possible that the upper layers of soil freeze and thaw as the ambient temperature changes; these layers are called the active zone. Below that, the soil may be permanently unfrozen or it could be permanently frozen, a leftover from the last ice age (called permafrost) (Figure 11.1). The length and severity of the cold season, the amount of snowfall, hydrogeological conditions, building density, and vegetative cover are among the main factors affecting the depth and duration of soil freezing.

Cold temperatures also affect the processes that take place in toilets and FSM systems, which in turn affect requirements for design, construction, and operation. In emergencies, extreme cold adds extra difficulties to the affected populations and those wishing to support them, including engineers. Some of the main effects are listed in Box 11.1.

There is a limited choice of feasible, cost-effective, proven systems for emergencies in cold regions that adequately protect human health and the environment. Many designs and processes that work well in tropical or temperate climates must be modified to work in cold climates. The required modifications can be prohibitively costly or complex, if they work at all. In general, the simpler the solution the better.

Some of the technologies described in the earlier chapters of this book may work in very cold climates provided they are adapted to suit the temperatures and social needs. This section describes the key impacts of cold weather on the different elements of the toilet and FSM chain.

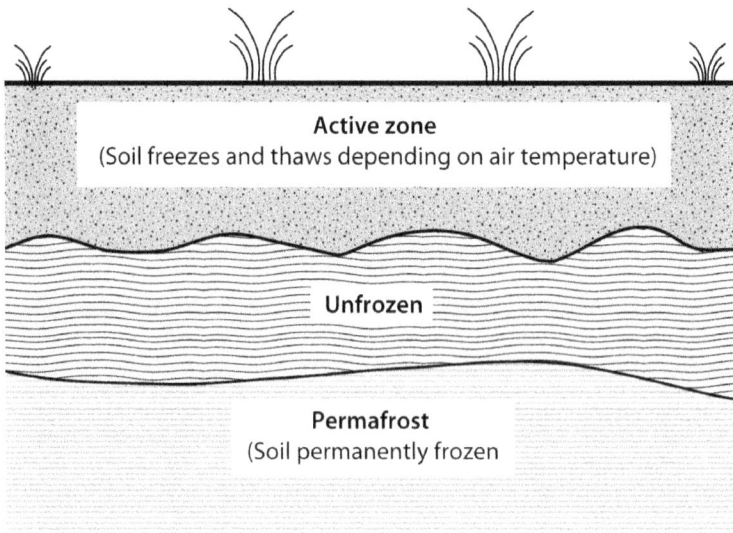

Figure 11.1 Cold climate soil zones

Box 11.1 Impact of extreme cold on sanitation provision

- Many biochemical treatment processes critical to protecting public health and the environment from pollution are less effective or do not work at all.
- Frozen saturated soil is impermeable, so any system that depends on infiltration of the liquid portion of the waste into the soil will not work.
- The waste in the pipes, pits, tanks, vaults, and other containers can freeze, thus blocking and potentially damaging them. Freezing liquids also expand, potentially causing major damage to containing structures. Thawing them is also very difficult during the cold season.
- Wastewater travelling in open channels can freeze and block them. It may also create an ice hazard on surrounding land surfaces. When the wastewater thaws, it can pollute the surrounding area.
- Soil movement caused by freezing and thawing of the water in the soil pores can damage facilities and infrastructure.
- The construction process is affected; digging in frozen soil is difficult if not impossible and concrete made in freezing conditions may fail to gain strength.
- Structures may need to be designed and built to resist loads from snow and ice.
- Vulnerable populations, including the young and the elderly, may experience increased difficulty in accessing and using toilet facilities.

11.1   User interface and containment

The effects of cold temperatures on the user interface depend on whether the facility uses a wet or dry interface. In cold regions, toilets with a water seal or a cistern for water for flushing must be in a heated building. When exposed to freezing temperatures, the water will freeze and expand, blocking the interface and damaging fixtures and pipes. Alternatively, pipes, fixtures,

Figure 11.2 Frozen excreta build-up in a pit toilet

tanks, and fittings must be insulated and equipped with heat tape or other means of keeping them from freezing and to thaw them in case of accidental freezing.

Dry interface toilets are generally better suited to cold regions, although excreta can freeze in pits. Frozen waste can form a solid pyramid below the place where the waste enters the pit, eventually blocking the waste entry point where the user defecates (as shown in Figure 11.2. One solution is to install a slot-shaped defecation hole. Users can defecate at different points along the slot, creating more of a ridge than a pyramid in the frozen excreta

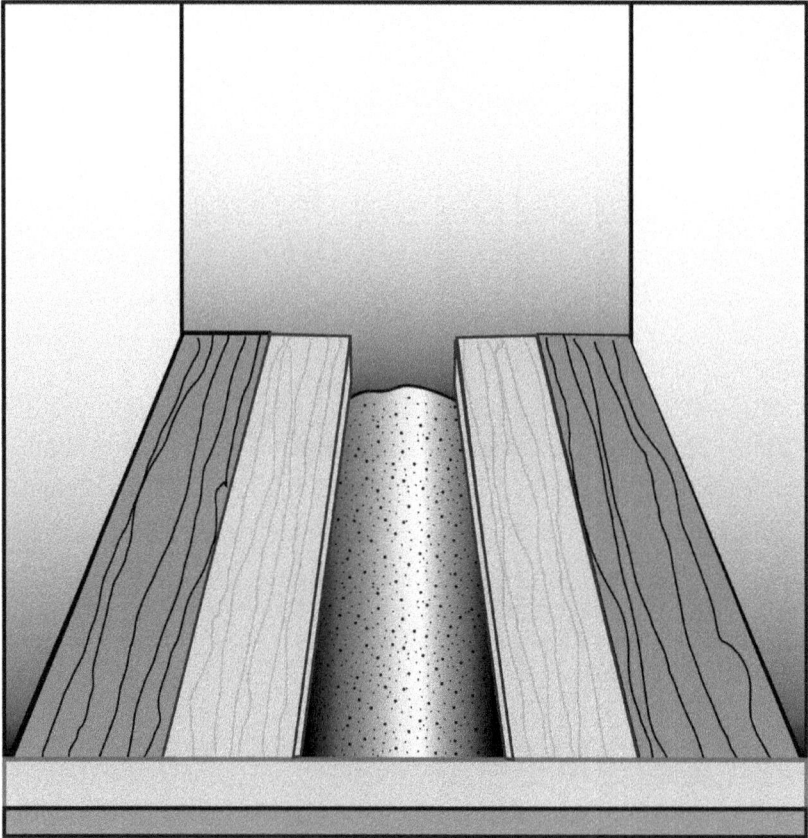

Figure 11.3 Pit defecation point for freezing excreta
Note: Pit defecation point spans the full width of the floor

(Figure 11.3). If a urine diversion interface is used, it must be carefully designed so that the urine does not freeze and block fixtures or pipes.

If users want a pedestal interface, materials must be chosen carefully. Very cold temperatures freeze human skin rapidly so appropriate, well-insulating materials are needed. Seats can be heated, but only where there is a reliable source of electrical power and people are willing to pay for it. Cloth covered seats such as shown in Figure 11.4 are difficult to keep hygienic and are therefore unsuitable for communal toilets.

11.2 Cubicle

Freestanding toilet cubicles should shelter users from the cold as much as possible. Walls should be substantial enough to retain some heat when in use. A roof is essential and, in some cases, snow loading and snow shedding

Figure 11.4 Toilet cubicle in a cold region

Note: The figure shows a 150 mm pipe for the urinal, cloth on the pedestal seat to protect against the cold, a bucket of dry sand for adding after defecation to absorb moisture and minimize odour, and a bucket for used anal cleaning materials.

may need to be considered. For wet interface options, the cubicle must be heated. Facilities for handwashing must also be insulated and any wastewater discharged away from the toilet to prevent icing of the surrounding area. Drainage pipes above ground must be insulated and may require heating to prevent freezing.

Communal toilet blocks under a single roof have some advantages over separate toilets. They are easier to heat and fittings (such as those for handwashing) can be protected from freezing more easily. Cubicles can be dedicated to individual family groups to prevent the issues of shared facilities.

11.3   Containment (container or pit)

Pits and vaults must be located either in or below a heated building, protected by insulation and/or an additional heat source, or extended below the soil active zone. Moreover, during the cold season, the air in pits, tanks, or vaults can conduct cold from the surface to the soil around the pit, causing that soil

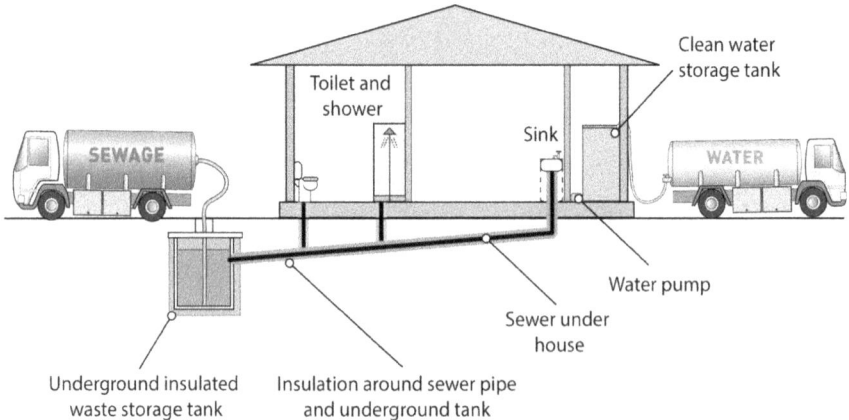

Figure 11.5 Household sewerage network with insulated cesspit

to freeze to a greater depth than the soil further away. Underground tanks are sometimes protected by a layer of insulation between the surface and the top of the tank to prevent this happening (Figure 11.5).

11.4   Emptying and collection

Liquid and solid waste stored within the soil active zone will accumulate in pits, tanks, vaults, or other containers during the cold season because the waste cannot be emptied when frozen and liquid cannot infiltrate into frozen soil. It is usually necessary to empty containers at the start of the cold season to ensure that there is enough empty space to store all the waste generated during the cold weather. Also, the reduction in volume of the faecal matter that occurs during decomposition will slow or stop during the cold season, so accumulation rates will be greater.

In the active zone, where soil freezes and thaws, pit linings, tanks, and vaults must be carefully designed and constructed to avoid damage from soil movement. This movement, in addition to the forces from freezing and thawing of the sludge inside, can exert tremendous pressure and crack or displace the walls. Freestanding containers must also use insulation and/or heating to make them resistant to the cold. Alternatively, they can be made of a semi flexible material such as polyethylene that can expand and contract without damage. When the warm season starts, drainage from melting ice and snow must be controlled so that the meltwater does not enter the waste system.

11.5   Sludge transport

Sludge transport, whether by vehicle or through sewer pipes, will be affected by the cold. Vehicles, equipment, and containers used to transport faecal sludge should be resistant to freezing and cold temperatures. With container-based systems, open trucks or trailers may be used to hold the containers. For liquid waste that is not frozen, tankers can be equipped with insulated tanks. Vehicles may need to be equipped with special tyres or chains for traction on icy roads. Also, water for washing vehicles can present an ice hazard if it spreads over the ground.

The contents of sewer pipes can freeze, blocking and damaging pipes and joints. To prevent freezing, pipes can be buried below the active zone (but above the permafrost) or routed through an utilidor (an insulated or heated conduit). Pipes passing through the active zone where the soil freezes and thaws or that are exposed to freezing air should be well insulated and may need to be equipped with heat tape to prevent freezing and to thaw them in case of accidental freezing. Flexible joints can help prevent damage from soil movement. Also, vents on pipes must be kept clear of ice and snow because the interface water seals can siphon dry if a vent freezes shut.

11.6   Treatment

Biological processes are heavily temperature-dependent, and they slow or stop in cold temperatures. Some pathogens and many helminth eggs and viruses can survive freezing temperatures for long periods. Overall, however, the effect of very low temperatures on decomposition and on the survival rate of pathogens in excreta is complex and not fully understood. Waste treatment plants serving large populations may continue to function because of the latent heat from the incoming wastes, but waste stabilization ponds generally freeze (except perhaps anaerobic ponds).

Because biological activity decreases in cold temperatures, physical and chemical methods may be required to stabilize excreta. However, many chemical and physical processes are also affected by cold temperatures, so they can be used for treatment only in the warm season. Neither lime nor urea react with ice, so they are unable to treat frozen sludge. Physical processes can also be affected because the viscosity of liquids increases as temperature decreases, affecting, among other factors, filtration and settlement rates. Treatment processes depending on live plant filtration such as constructed wetlands and planted drying beds stop working because of freezing of the filter matrix and the dormancy of the filter plants.

11.7   End use and disposal

Generally, agriculture is limited when the ground is frozen so there is no demand for fertilizer. In any case, composting processes cease at low temperatures unless undertaken in a heated space. In principle, dried and liquid treated sludge can be spread on the surface of agricultural land if they are not frozen (see Chapter 10, section 6.3.

Biogas plants do not function well in cold temperatures because the biochemical processes that generate the gas cease to function, and so they cannot be recommended where the temperature in the unit drops below 5°C. In cold regions where fuel is scarce or costly, dried treated sludge could be processed to provide a fuel for households to burn for heat, but the health and environmental issues may be considerable; this option should only be undertaken after extensive investigation (see Chapter 10, section 6.6). In some cases, large holding ponds are created to store effluent during the cold season for discharge when the temperature increases. Chapter adapted from Leblanc et al., 2019.

CHAPTER 12

Faecal sludge management during health epidemics

During any emergency, faeces are possibly the greatest risk to human public health due to their incredibly high pathogen loading. Excreta management interventions must be given the highest priority as they have the potential to prevent health epidemics. This chapter looks at three specific health epidemics, but the measures described apply to all faecal–oral-related epidemics.

12.1 Cholera

Cholera is a faecal–oral disease that causes infection of the small intestine leading to severe watery diarrhoea, rapid dehydration, and death if left untreated. Cholera occurs in humanitarian emergencies and in settings where outbreaks occur regularly among the same populations, usually coinciding with the rainy season. While cholera can spread quickly through the environment, there are several known and effective ways to halt transmission. Practices that isolate faeces from food and water – such as treating and storing water safely, improved toilet facilities, and good handwashing practice using soap – and oral rehydration therapy are essential to control a cholera outbreak (Section adapted from WHO and UNICEF, 2014).

12.1.1 Excreta disposal interventions

In cholera epidemics and risk-prone areas, safe excreta collection and disposal in household, community, and medical facilities are critical. Box 12.1 lists the most important aspects of safe FSM. (Source ** Gensch, et al., 2018).

Box 12.1 Focus areas for safe faecal sludge management in cholera epidemics

- **Improved access to and use of safe excreta disposal:** Faecal matter contains and spreads cholera bacteria and it must be kept away from water and food (containment). All food and water that could be contaminated must be disinfected prior to consumption (disinfection). Suspected or confirmed cholera cases must be provided with separate toilets that are not used by other individuals. Adequate numbers of functioning, accessible, appropriate, and safe toilets must be provided for staff, patients, and caregivers (Table 5.2) and these must not contaminate the healthcare setting or water supplies. More information on the treatment of cholera faeces is given in Box 12.2. The layout of a typical cholera treatment centre, including the location of key human waste management services, is shown in Figure 12.1.

(Continued)

Box 12.1 Continued

- **Environment free from human excreta:** All toilets must be provided with functional and clean handwashing facilities with soap. All people, including children, must be prevented from practising open defecation and all faeces must be disposed of safely in a toilet or buried. Excreta disposal facilities need to be provided in markets, public places, and institutions, along with functioning and well-managed handwashing facilities. They should be culturally appropriate and a sustainable cleaning and management system should be established for public and communal facilities.
- **Handwashing:** Effective handwashing practices must be explained, promoted, and supported within the whole of the affected population, especially among healthcare providers.
- **Personal protective equipment:** PPE (e.g. boots, masks, gloves, clothing, etc.) must be provided for those involved in the operation and maintenance of toilets and the faecal sludge management service chain.
- **Disposal of faeces from bed-ridden patients:** Bed-ridden patients will defecate into containers that should be emptied as soon as possible after use to reduce odour and accidental spillage. Faeces can be emptied into an existing toilet – or, better still, into a specially designed sluice (Figure 12.2). Where possible surround the sluice by a low wall to prevent splashing and contamination of the users. The sluice discharges via a water seal to a pit or sealed tank, depending on soil conditions and the possibility of groundwater contamination. The container must be washed out in the sluice and disinfected before being returned to use.
- **Chlorine solution disinfection:** Different chlorine solutions (with different percentages of free residual chlorine) must be available for different purposes:
- 0.05 per cent for handwashing with soap, skin disinfection, laundry (patient and administrative), toilets, kitchen, mortuaries, and waste areas (or alternatively alcohol-based hand rub);
- 0.2 per cent for disinfecting floors, objects, beds, clothes, and kitchen utensils of patients;
- 2.0 per cent to add to excreta/vomit for disinfection, to wash dead bodies, and for footbaths at the entrance to clinical controlled areas and at the exit from toilets (or alternatively lime treatment).

Additional information can be found in (UNICEF, 2017)

Box 12.2 Safe disposal and treatment of cholera vibrio

The cholera bacteria (or vibrio) is a complex organism, making it difficult to decide how to treat faeces containing them. Based on the information below, the treatment options are likely to be infiltration, combined storage and dehydration using evaporation to prevent the need for treatment of the liquid effluent, or disinfection. Whichever method is chosen, a minimum storage time of two years is recommended.

The lifecycle of cholera bacteria is complex and only rarely linked to humans. They survive mainly in an environment that is a mixture of fresh and salt water and they can remain viable for a long time. For some reason, not fully explained, when some vibrio come into contact with the human gut, they are changed and become dangerous and highly infective. It is these bacteria that have the potential to create an epidemic. As far as we know, these epidemic bacteria can remain infective in faeces for up to 50 days, and in a very dry environment such as the soil surface for around 16 days. However, the bacteria survive well in water and may remain viable in shellfish, algae, or plankton in coastal regions.

Sources: (Jubair, et al., 2012), (Center for Food Security and Public Health, 2003)

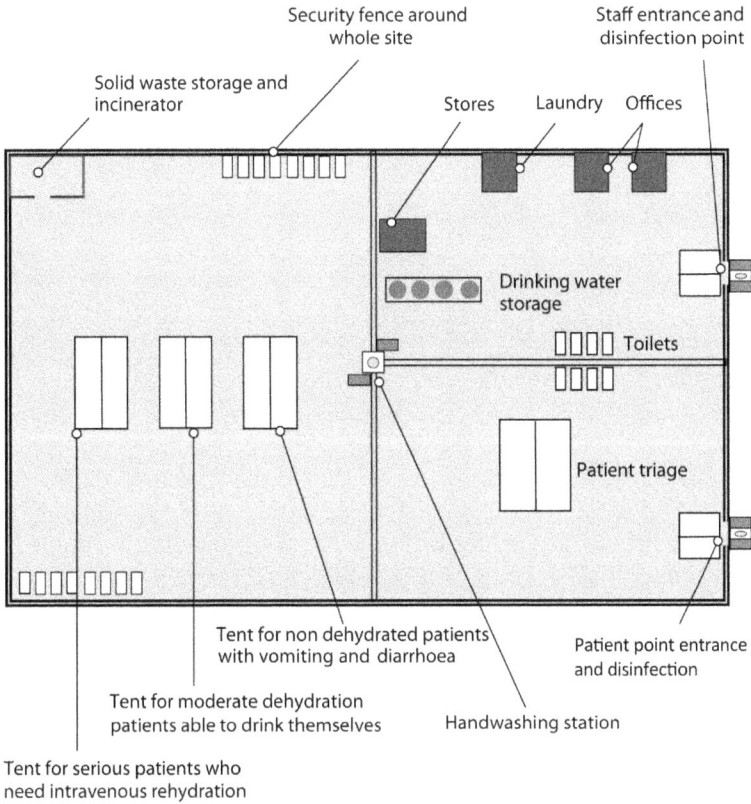

Figure 12.1 Emergency cholera treatment centre
Source: Adapted from (Médecins Sans Frontières, n.d.)

Figure 12.2 Simple sluice for disposal of faeces

12.2   Ebola

Existing recommended human waste management and hygiene measures in healthcare settings are important for providing adequate care to patients and for protecting patients, staff, and carers from infection risks. Of particular importance are the following actions: keeping excreta (faeces and urine) separated from drinking-water sources; handwashing with soap; and containment of excreta so that they are effectively separated from human contact.

12.2.1 Toilets in health facilities

Suspected or confirmed Ebola cases should be provided with separate toilets that are not used by other individuals. If flush toilets are used, it is important that standard procedures for wastewater treatment are followed, including, at a minimum, on-site septic tank treatment with controlled sludge removal for further treatment. Containing the wastewater for a period of time not less than a week (Bibby, 2015) prior to secondary biological treatment will allow for natural die-off of the Ebola virus and will significantly reduce the concentration of Ebola virus, along with other pathogens that may be found in the wastewater.

If healthcare facilities are connected to sewers, a risk assessment should be conducted to confirm that wastewater is contained within the system (i.e. that it does not leak) prior to its arrival at a functioning treatment and/or disposal site. Risks pertaining to the adequacy of the collection system, or to treatment and disposal methods, should be assessed according to a 'safety planning' approach, with critical control points prioritized for mitigation.

For smaller facilities, if space and local conditions allow, pit toilets may be the preferred option. Alternatively, faeces may be collected and stored in containers (that can be sealed) for at least a week before being disposed of in a sanitary facility. Urine should be collected separately and disposed of in a leach pit. Standard precautions should be taken to prevent contamination of the environment by faeces and urine. These precautions include ensuring that there is a distance of at least 1.5 metres between the bottom of the pit and the groundwater table (more in coarse sands, gravels, and fissured formations) and that the toilets are located at least 30 metres horizontally from any groundwater source (including both shallow wells and boreholes). Figure 12.3 shows the layout of an Ebola hospital recommended by the WHO.

12.2.2 Handling and treatment of faeces and urine in health facilities

The key to controlling the hazards associated with the presence of the virus in the body fluids of infected individuals lies in the rigorous enforcement of protocols to separate and contain **all** body fluids (including faeces and urine). Faeces from suspected or confirmed Ebola cases must be treated as a biohazard and handled as little as possible. All direct human contact

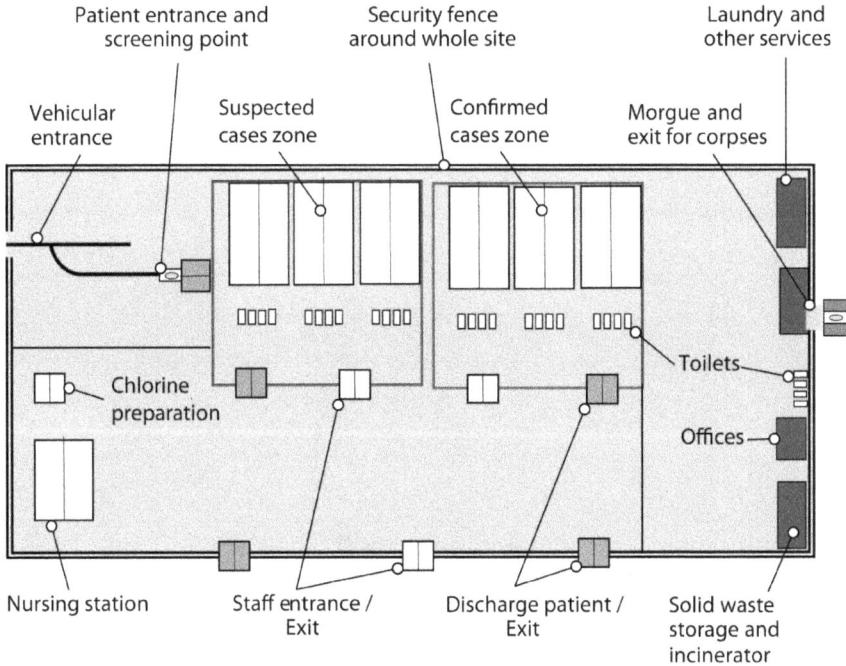

Figure 12.3 Emergency Ebola hospital layout
Source: Adapted from (McCartan, 2015)

with excreta should be avoided and full PPE should be worn by all workers handling faeces (Figure 12.4). Such equipment includes heavy-duty rubber gloves, impermeable gown, impermeable apron, closed shoes (e.g. boots), facial protection (mask and goggle or face shield), and ideally a head cover. Workers should be properly trained in putting on, using, and removing PPE so that these protective barriers are maintained and not breached.

If the patient is unable to use a toilet, excreta should be collected in a clean bedpan and immediately and carefully disposed of into a separate toilet used only by Ebola cases or suspected cases (see details above for toilets). After collection and disposal of the excreta from the bedpan, the bedpan should be rinsed with 0.5 per cent chlorine solution to disinfect the pan, disposing of the rinse water in drains or a toilet. Depending on the dirtiness of the pan, it may need to be rinsed twice.

If it is not possible to dispose of the excreta immediately, the following procedure can be used to accelerate the inactivation of the Ebola virus and to contain the faeces temporarily:

- In a 10-litre covered bucket, first add approximately 600 ml of a 10 per cent solution of hydrated lime (i.e. 100 g of calcium hydroxide powder in 1 litre of water).

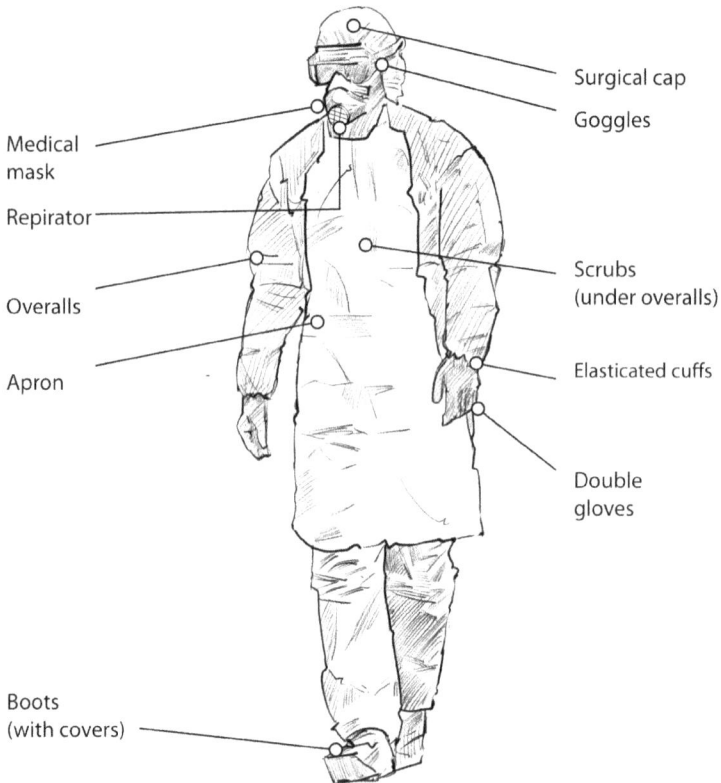

Figure 12.4 Full PPE for Ebola epidemics

- Carefully add the excreta from the bedpan into the bucket, leaving sufficient space in the bucket to add safely at least an additional 400 ml of lime slurry if necessary.
- Rinse and disinfect the bedpan as described above.

The final product should continue to be treated with caution and be carefully disposed of in a toilet by a person wearing full PPE. If excreta are on surfaces (linens, floor, etc.) they should be carefully removed and immediately disposed of in a toilet. If this is not possible immediately, temporary containment using a bucket and lime as detailed above is recommended. All surfaces in contact with excreta should be disinfected (see details below). Chlorine is not suitable for disinfecting containers containing large amounts of solid and dissolved organic matter.

12.2.3 Emptying toilets and septic tanks and off-site transport

Septic or holding tanks should be designed to hold wastewater for as long as possible with a regular emptying schedule. Full PPE should be worn at all times when handling or transporting excreta off-site and great care should be taken to avoid splashing. For crews, this includes pumping out tanks or unloading pumper trucks. After handling, and once there is no risk of further exposure, individuals should safely remove PPE before entering the transport vehicle. All equipment must be fully cleaned and disinfected after use.

12.2.4 Wastewater treatment

There is no evidence to date that Ebola has been transmitted via sewerage systems, with or without wastewater treatment. Sewerage should ideally be treated in well-designed and well-managed centralized wastewater treatment works. Waste stabilization ponds are particularly well suited to the destruction of pathogens, as relatively long retention times (20 days or more) combined with sunlight, elevated pH levels, and other factors serve to accelerate pathogen destruction.

12.2.5 Disposal of grey water

Current WHO recommendations advise to use chlorinated water (0.5 per cent) to wash any reusable PPE (all disposable items should **not** be reused but should be disposed of safely), as well as surfaces that may have come into contact with bodily fluids. This concentration of chlorine is sufficient to inactivate the Ebola virus in water that is relatively free of solids (less than 10 mg/litre). As such, this grey water, which has already been chlorinated, does not need to be chlorinated or treated again. It is important, however, that such water is disposed of in drains connected to a septic system or sewer or into a soakaway pit. If grey water is disposed of into a soakaway pit, the pit should be fenced off within the health facility grounds to prevent tampering and to avoid possible exposure in the case of overflow (Section adapted from WHO and UNICEF, 2014).

12.3   Coronaviruses

Coronaviruses are a group of related RNA viruses that cause diseases in mammals and birds. In humans and birds (and possibly other animals), they cause respiratory tract infections that can range from mild to lethal. Mild illnesses in humans include some cases of the common cold (which is also caused by other viruses, predominantly rhinoviruses), while more lethal varieties can cause SARS, MERS, and COVID-19. At the time of writing there is no indication that COVID-19 can persist in drinking water. In wastewater, some recent studies have found RNA fragments but not infectious virus present. The makeup of the COVID-19 virus is similar to that of other coronaviruses, for which there are data both on their survival in the environment and on effective measures to inactivate them.

Although little evidence is available, some data suggest that transmission via faeces is possible but unlikely, especially when small particles of faeces become suspended in the air. Because of the potential infectious disease risks from excreta, including the potential presence of SARS-CoV-2, wastewater and sludge should be contained and either treated on-site or conveyed off-site and treated in well-designed and managed wastewater or faecal sludge treatment plants. Each stage of treatment, which combines physical, biological, and chemical processes (e.g. retention time, dilution, oxidation, sunlight, elevated pH, and biological activity), results in a further reduction of the potential risk of exposure and accelerates pathogen reduction. A final disinfection step may be considered if existing treatment plants are not optimized to remove viruses.

Workers should follow standard operating procedures, including wearing appropriate PPE (Table 12.1), minimizing spills, washing dedicated tools and clothing, performing hand hygiene frequently, obtaining vaccinations for sanitation-related diseases, and self-monitoring for any signs of coronavirus or other infectious disease, with the support of their employer. Additional precautions to prevent transmission between workers, which apply to the general population as well, include avoiding touching the eyes, nose, or mouth with unwashed hands, sneezing into one's sleeve or a disposal tissue, practising physical distancing while working and travelling to and from work, and staying home if one develops symptoms associated with coronavirus (e.g. fever, dry cough, fatigue).

Table 12.1 A guide to safe PPE for manual workers

Protection	General contact with confirmed or possible COVID 19 cases	Aerosol Generating Procedures of High Risk Areas
Eyes	Eye protection to be worn based on risk assessment	Eye protection, eye shield, goggles or visor
Mouth & Nose	Wear fluid resistant surgical mask	Wear filtering facepiece respirator
Body	Wear disposable apron	Wear long sleeved fluid repellent gown
Hands	Gloves	Gloves
Hand hygiene	Clean hands before and after patient contact and after removing some or all of your PPE.	
Equipment hygiene	Clean all the equipment that you are using according to local policies.	
Equipment selection	Use the appropriate equipment for the situation you are working in.	
PPE removal	Take off all your PPE safely	
Personal care	Take breaks and hydrate yourself regularly	

Source: Adapted from (UK Government, n.d.).

12.3.1 Management in healthcare settings

People with suspected or confirmed SARS-CoV-2 or other coronavirus infections should be provided with their own toilet (either a flush or dry toilet). Where this is not possible, patients sharing the same ward should have access to toilets that are not used by patients in other wards. Each toilet cubicle should have a door that closes. Flush toilets should operate properly and have functioning drain traps. The toilet should be flushed with the lid down to prevent droplet splatter and aerosol clouds. If it is not possible to provide separate toilets for coronavirus patients, the toilets they share with non-coronavirus patients should be cleaned and disinfected more regularly (e.g. at least twice daily by a trained cleaner wearing PPE – an impermeable gown (or, if not available, an apron), heavy-duty gloves, boots, mask, and googles or a face shield). Healthcare staff should have access to toilet facilities that are separate from those used by patients.

Faulty plumbing and a poorly designed air ventilation system were among the contributing factors for the spread of the aerosolized SARS-CoV-1 coronavirus in a high-rise apartment building in Hong Kong. If healthcare facilities are connected to sewers, a risk assessment should be conducted to confirm whether wastewater is contained and does not leak from the system before it reaches a functioning treatment and disposal site.

If healthcare facility toilets are not connected to sewers, hygienic on-site containment and treatment systems should be provided, such as pit toilets and septic tanks. Sludge should be safely contained. When containers are full, they should be transported off-site for treatment or treated on-site if space and soil conditions permit. For unlined pits, standard precautions should be taken to prevent contamination of the environment. There is no reason to empty toilet pits and holding tanks of excreta from suspected or confirmed coronavirus cases unless they are at capacity. In general, the best practices for the safe management of excreta should be followed. For personnel working with untreated sewage, for whom there are considerable risks of infection, standard PPE should be worn. Untreated faecal sludge and wastewater from health facilities should never be released on land used for food production or in water used for aquaculture or disposed of in recreational waters.

It is critical to perform hand hygiene when there is suspected or known contact with faeces. If the patient is unable to use a toilet, excreta should be collected in either a diaper or a clean bedpan and immediately disposed of carefully into a separate toilet or pit toilet used only by suspected or confirmed coronavirus cases. In all healthcare settings, including those with suspected or confirmed coronavirus cases, faeces must be treated as a biohazard. After disposing of excreta, bedpans should be cleaned with a neutral detergent and water, disinfected with a 0.5 per cent chlorine solution, and then rinsed with clean water. The rinse water should be disposed of in a drain or toilet. Chlorine is not effective for disinfecting faecal waste containing large amounts of solid and dissolved organic matter. Accordingly, it is not recommended to add chlorine solution to fresh excreta. It is also possible that such addition can introduce risks associated with splashing. Used diapers should be sealed in a waterproof plastic bag and sent for incineration or burial. See Chapter 12, section 2 for recommendations on the cleaning of multi-use PPE.

If grey water includes household bleach used in prior cleaning, it does not need to be chlorinated or treated again. Likewise, used bathing water from coronavirus patients does not need to be disinfected. However, it is important that such water is disposed of in drains connected to a septic system or a sewer or in a soakaway pit.

12.3.2 Human waste management practices in communities

When there are suspected or confirmed cases of coronavirus in the home setting, immediate action must be taken to protect caregivers and other family members from the risk of contact with respiratory secretions and excreta that may contain coronavirus. Support must include clear instructions on the safe and correct use and storage of cleaning products and household bleach, including keeping them out of reach of children to prevent harm from misuse, including poisoning. Frequently touched surfaces throughout the patient's care area should be cleaned regularly, such as tables and other bedroom furniture. Cutlery and crockery should be washed and dried after each use and not shared with others. Bathrooms shared by coronavirus patients and other household members should be cleaned and disinfected at least once a day. Regular household soap or detergent should be used for cleaning first and then, after rinsing, regular household bleach containing 0.1 per cent sodium hypochlorite (i.e. equivalent to 1,000 ppm or 1 part household bleach with 5 per cent sodium hypochlorite to 50 parts water) should be applied. PPE should be worn while cleaning – including mask, goggles, a fluid-resistant apron, and gloves – and hand hygiene should be performed after removing PPE. Where households have limited resources, efforts should be made to provide PPE supplies – at a minimum, masks – and hand hygiene supplies to households caring for coronavirus patients. Consideration should be given to safely managing the whole human waste management chain, starting with ensuring access to regularly cleaned, accessible, and functioning toilets and to the safe containment, conveyance, treatment, and eventual disposal of sewage and sludge.

The importance of disseminating information to the general public and caregivers cannot be over emphasized. The containment of coronaviruses is predominantly down to the practices of the individual. Section adapted from WHO and UNICEF 2020.

CHAPTER 13
Technology selection

Designing an integrated faecal sludge management (FSM) chain is highly complex and requires a large amount of information. However, there is rarely the time or the information available during the early stages of an emergency to carry out such a rigorous design. A highly flexible approach to selection must be taken to allow for rapid changes in the scenario and a gradual increase in knowledge. In general, the priority at the beginning of an emergency is to construct toilets; development of the remainder of the FSM chain tends to follow later as the situation develops. Even the decisions on toilet design are complex, requiring consideration of multiple technical, social, institutional, environmental, and financial constraints, as suggested in Box 13.1. Even when all the available data has been collected, there's still the issue of how to use it to make decisions. Over the years, numerous organizations have attempted to develop a decision-making guide for selection, but at best they have been limited to one particular aspect of the FSM chain. In recent years the problem has received more attention, and although at the time of writing nothing currently exists, it may do so soon. To keep up to date with progress on this subject, you should consult the ELRHA website (ELRHA, n.d.).

In practice, design is usually based on a good understanding of the situation, experience of working in similar situations, and consultation with other experienced personnel. In major emergencies this approach tends to work well because of the involvement of multiple international organizations, but in smaller emergencies, where responsibility is given to those who happen to be available locally, it can be very difficult. As a general rule, follow the steps in the FSM chain:

- Select the toilet technology preferably based on what users used prior to the emergency.
- If the toilet includes some form of containment, decide how it will be emptied and transported.
- If the toilet is a water-based system, decide where the wastes will be carried to and where the water will come from then:
  - Select the treatment process.
  - Select the methods for ultimate disposal/reuse.

This is not a linear process: decisions made further down the chain may impact on earlier decisions. The process is iterative until an overall design is reached (Figure 13.1).

Box 13.1 The principles of emergency toilet selection

Determine which technological options could work based on:

- Cultural considerations
- Sanitary practices of the affected populations at their location of origin
- Beneficiary needs and preferences
- Anal cleansing practices
- Willingness to share toilets
- Cultural taboos
- Availability of water for flushing and handwashing
- Environmental conditions
- Hardness/rockiness of the soils
- Soil infiltration rates
- Depth to groundwater table
- Whether groundwater is potable or non-potable (saline >1,500 μScm^2)
- Population density (rural/urban/peri-urban setting)
- Available space for toilet construction
- Permission from the landowner to build toilets
- Availability of toilet construction materials
- The presence of existing FSM infrastructure such as sewer lines or treatment plants
- National legislation concerning FSM and/or environmental pollution
- For each option consider the financial, environmental, technical, health, institutional, implementation speed, minimum standards, and social implications of its use
- Prepare a shortlist of possible options
- Determine the impact of each shortlisted option on the rest of the FSM chain
- Discuss the final shortlisted options with partner institutions, emergency planning authorities, local government, and future users to determine best options
- Review frequently as circumstances change and additional information becomes available

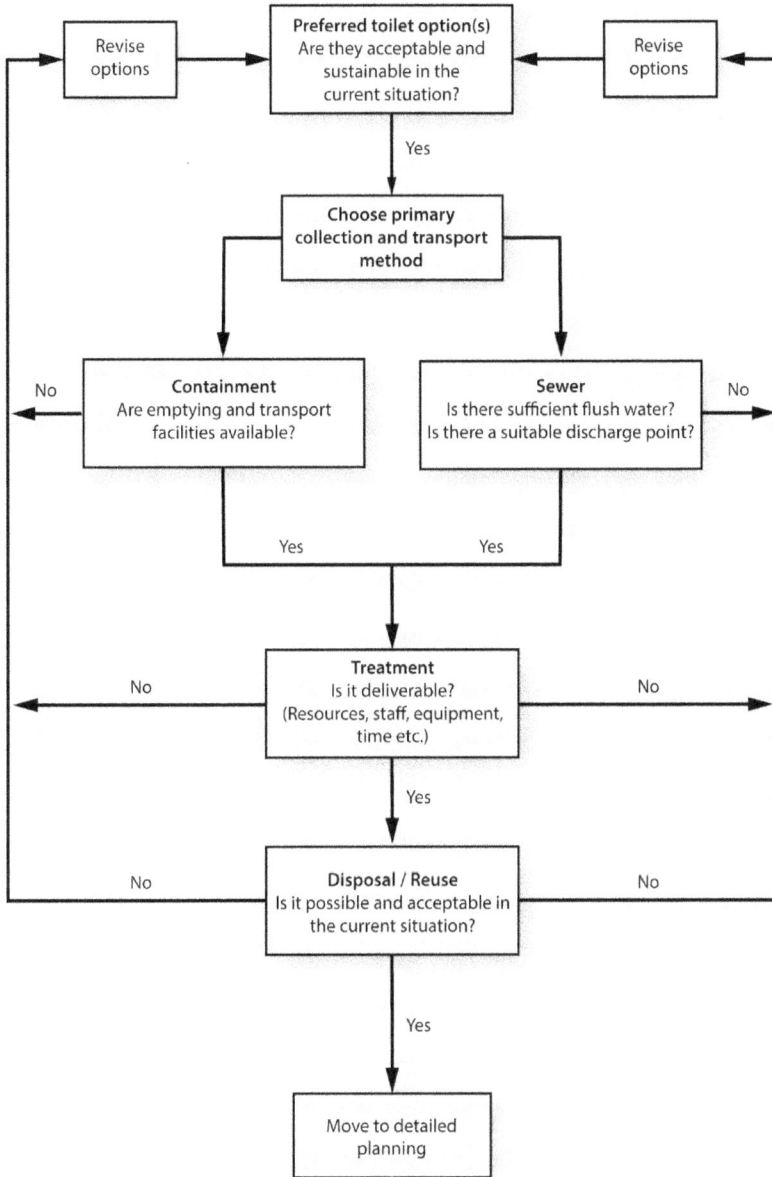

Figure 13.1 Outline of the FSM selection process

CHAPTER 14
Emergency sullage/grey water disposal

14.1 Introduction

Sullage, sometimes called grey water (because of its colour), consists of the liquid wastes from domestic, communal, institutional, and commercial activities. Sullage does not include excreta-related liquids, which are sometimes known as black water. Indiscriminate disposal of sullage is a common sight during an emergency. Spilt water from communal water points, drainage from laundry sites, and runoff from bathing points are all common in damaged urban areas and displaced people's camps. But with a little effort and forethought, most of it can be avoided.

Surface and storm water drainage is a bigger issue and encompasses the movement of all surface water around the affected area. This is closely linked to site layout and management. Surface and storm water drainage is not covered by this book (further information can be found in (ELHRA, 2019)).

14.2 Risks from sullage

Although not such an obvious health risk as excreta, sullage poses several indirect risks that should be considered. These include:

- possible inclusion of traces of excreta (including pathogens) from soiled clothing and bathing;
- providing breeding sites for water-related insect vectors;
- soil erosion around temporary shelters;
- filling of pit toilets and solid waste pits;
- pollution of surface and groundwater;
- obstruction of access paths and walkways;
- reduced morale from living in an unsanitary and contaminated environment.

14.3 Sources and types of sullage

The most common sources in emergency settings are listed in Box 14.1. In most temporary settlements, water is carried to the dwelling from communal water points, limiting the volumes of sullage generated in the home. Even so, people should be advised on the most appropriate way of disposing of their sullage. People affected by an urban emergency may be living in their own homes and will tend to use the disposal systems they used previously. If those disposal

---

Box 14.1 Common sources of sullage production in emergency settings

water points;
bathing areas;
kitchens or feeding centres;
laundry slabs or areas;
market areas (washing of market stalls, abattoirs, etc.);
small-scale commercial activities such as restaurants;
healthcare centres.

---

systems have been badly damaged or people have been forced to relocate, discussions should be held with the community to decide on alternative disposal methods.

Where communal water points are used, the quantity of spillage can vary greatly. Installing good-quality taps, setting up water point management (24-hour access versus set times in the day for distribution and collection), and constructing dedicated laundry, bathing, and food preparation areas not only save water but considerably reduce the environmental impact from sullage. If the tap stand is used only for water collection, then the sullage will be clean, which simplifies disposal. Sullage generation from communal laundry, bathing, food preparation, or pot-washing facilities will generally be much more contaminated than from water points as illustrated in Table 14.1. In general, they have higher turbidity and levels of dissolved and suspended solids. If left untreated it will biodegrade and become odorous, a breeding ground for flies and mosquitoes, and attractive to scavenging animals and vermin. Again, the design of the area, the use of high-quality fittings, and good site management can have a significant effect on the volumes of sullage generated. The design and construction of water points and bathing and laundry areas are not covered by this book, further information can be found in (** Davis & Lambert, 2002).

Table 14.1 Typical sullage contamination from various sources

Source	Contamination
Kitchens	Cooked and uncooked animal and vegetable food waste, oils and fats, soap, silt and grit
Laundry	Laundry soap, silt and grit, oil, faeces, blood, urine
Bathing	Bathing soap, faeces, silt and grit, blood, urine
Healthcare centres	All of the above, depending on the type of facility, plus other non-faecal transmissible pathogens
Markets	Vegetable and animal products, silt and grit
Restaurants	Similar to kitchens

Note: The amount of faeces, blood, and urine in laundry and bathing sullage is usually very low but it can be significant from healthcare centres.

14.4   Selection criteria

When deciding on the most appropriate sullage disposal method to use, there are a number of factors to consider.

14.4.1 Ground conditions

Many treatment and disposal systems include infiltrating the sullage into the ground. Table 4.3 suggests infiltration rates for liquids with low levels of suspended solids in different soil types. Infiltration rates will be considerably lower where the quantity of suspended solids is high or the liquid contains fat. It is always advisable to undertake a percolation test (Appendix 1) prior to design and construction, but in the absence of more accurate data it is suggested that infiltration rates given in Table 4.3 are reduced by 50 per cent.

14.4.2 Water sources

In all cases it is a priority to prevent contamination of drinking water sources from sullage. The location of all existing or potential water sources should be mapped before locating and designing sullage generation points. Where sullage is discharged directly or indirectly into surface water, it is important that it is downstream of any water supply intakes. This will protect the source from cross-contamination and reduce the need for increased water treatment. It is also important to consider other downstream water uses such as irrigation or livestock management and what the effects of effluent discharge might have on them.

14.4.3 Water table

The height of the water table also affects subsurface infiltration and the risk of groundwater contamination, as described in Chapter 4, section 4.

14.4.4 Topography

The topography of the affected site will be a key factor in determining whether surface drainage techniques can be adopted. It is rare to find a site that is completely flat, although where this is the case, or nearly so, surface drainage becomes almost impossible. In general, a minimum gradient of 1:200 is recommended for the transport of sullage in earth drainage ditches.

14.4.5 Sullage quantity

The volume of sullage generated will also influence the technology choice made. Where only small quantities of sullage are generated, infiltration

Table 14.2 Typical sullage volumes

Institution	Sullage volume
Field hospital	40–60 litres/patient/day
Hospital with operating theatres	100 litres/intervention
Out-patient clinics	5 litres/patient/day
Cholera treatment centre	50 litres/patient/day
	10 litres/carer/day
Viral haemorrhagic fever centre	300–400 litres/patient/day
Feeding centre	25 litres/person/day
	10 litres/carer/day
Public bathing area – piped water provided	100 litres/user^
– no piped water provided	20 litres/user
Public laundry area – piped water provided	100 litres/user^
– no piped water provided	20 litres
Public water points	5–20 litres/user^
Household sullage based on daily per person water supply of 7.5 litres	2–5 litres
15 litres	6–12 litres
30 litres	10–20 litres
50 litres	15–30 litres

Note: ^These numbers vary widely depending on the quality of the water supply control mechanism (e.g. tap design, shower head, etc.) and how well the facilities are managed. Local assessment is always recommended.

Source: Adapted from (** Sphere Association, 2018)

may be appropriate even in low permeability soils. It may even be possible to use evaporation for disposal. Where large volumes are involved, disposal systems must be selected and sized accordingly. Existing systems may become inappropriate if water use increases greatly and they will need upgrading or replacing. Guidelines on sullage generation rates for public institutions are shown in Table 14.2.

14.4.6 Climate

In hot, dry climates, the use of evaporation or irrigation for sullage disposal may be viable, but in wetter climates the volume of rainfall must be considered and may even be used to dilute the sullage. In colder climates, the possibility of the drainage network freezing should not be overlooked

14.4.7 Socio-cultural considerations

Although sullage management is generally a less sensitive issue than excreta disposal or hygiene promotion, socio-cultural aspects should be considered.

Where sullage drainage channels pass through residential areas it may tempt people to use the sullage for domestic purposes. Cultural practices and traditions in terms of water use may also influence the volumes of water used and sullage generated. This may also affect when sullage is produced: for example, if large numbers of people bathe or do laundry at a particular time of day. Communities that have historically had access to large quantities of clean water will find it difficult to adapt to low water use and will tend to generate large quantities of sullage.

14.4.8 Sullage transport

Where possible, sullage should be transported in sewer pipes (see Chapter 8, section 4 for advice on simple sewer design); pipes isolate the sullage from the environment and prevent it from being contaminated by garbage or silt. Oils, fat, grease, and soap should be removed from the sullage prior to it entering the sewer to prevent it becoming blocked. The entrance to the sewer should be covered by a coarse screen to keep out gross solids. In general, the flow will be quite low and the size of pipe will be governed by availability and ease of maintenance. A minimum size of 75 mm diameter is recommended.

Carrying sullage in open channels, especially unlined channels, is not recommended; they require constant cleaning and maintenance and are a trip hazard. If they must be used, they should be as short as possible and preferably lined with bricks and/or concrete. Where possible, channels should be covered. In most cases, the size of channel necessary to carry the flow is very small, so select a size that is easy to construct, robust, and easy to clean. A channel with internal dimensions of 75–100 mm wide and 50–75 mm deep is often sufficient.

14.5 Sullage treatment

14.5.1 Gross solids removal

Where sullage has a high solids, oil, or soap content it will be necessary to separate these components prior to disposal. Sullage from large kitchens, feeding centres, laundries, and possibly medical facilities is most likely to require treatment prior to disposal. Inorganic solids such as silt and garbage will block water courses and soak pits and contaminate the environment. Organic solids such as vegetable matter will decompose, blocking infiltration systems, encouraging fly breeding and scavenging animals, and creating unpleasant odours. In many cases, a simple coarse filter made of loosely woven sacking or a slotted bucket or even a kitchen strainer will be sufficient provided it is cleaned regularly (Figure 14.1).

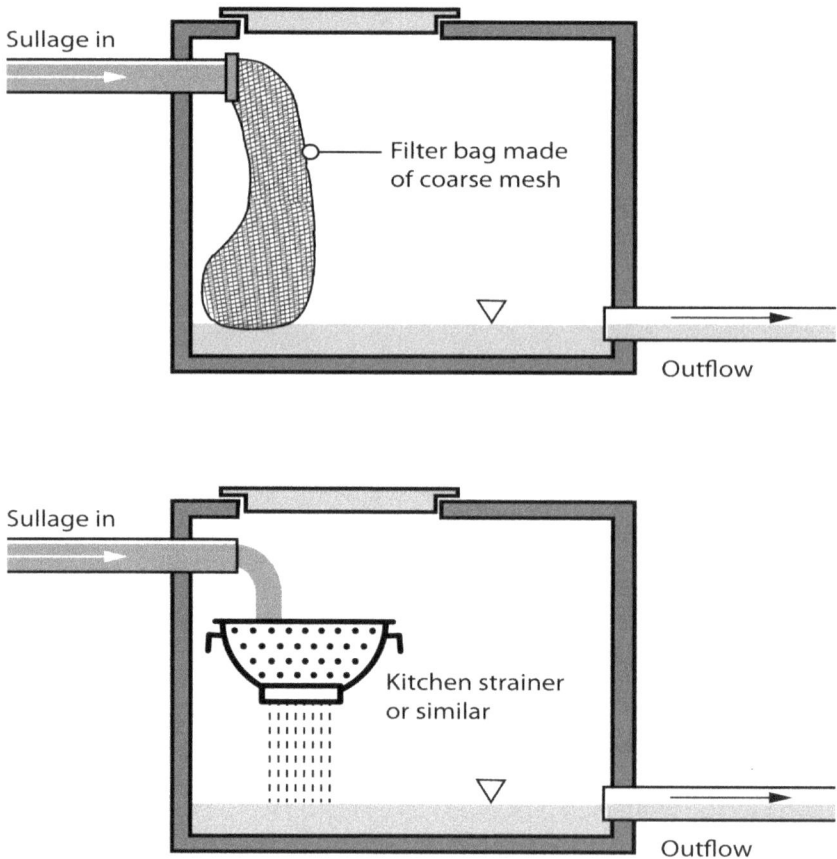

Figure 14.1 Coarse sullage filters

Advantages
- Simple design and easy to construct.

Disadvantages
- Requires very regular cleaning otherwise they quickly overflow or back up.
- Only suitable for small sullage flows – for larger flows, a larger coarse filter similar to those used for the primary filtration of faecal sludge can be used (see Chapter 9, section 2).

14.5.2 Grease traps

In general, soap and grease are held in solution in fresh sullage by the warm water temperature. As the water temperature drops, they come out of solution and can stick to nearby surfaces, blocking pipes and clogging infiltration systems. Therefore, grease traps should be located as close as possible to the point where the sullage is generated. A simple grease trap (Figure 14.2) consists

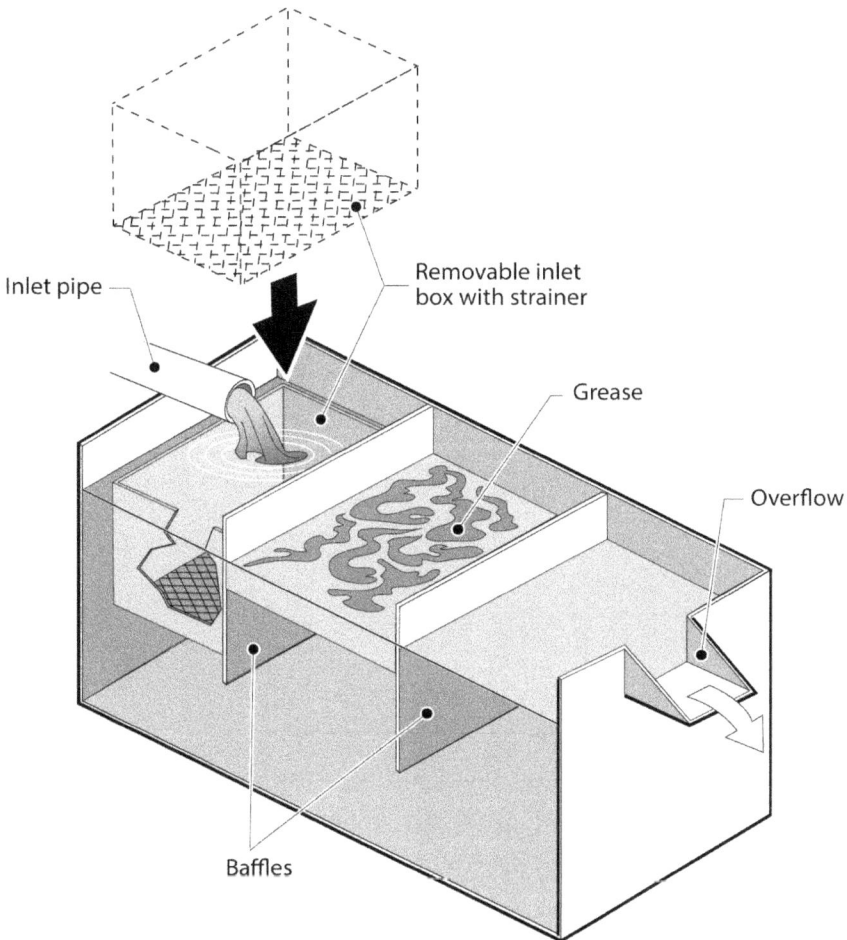

Figure 14.2 Grease trap

of a single tank containing an inlet with a strainer below to remove gross solids; this is followed by a series of baffles that allow liquid flow below them. Grease floats to the surface as the liquid cools and is captured by the baffle plates while silt settles to the bottom. Clarified liquid overflows at the outlet for further treatment or disposal. Grease traps should be cleaned regularly, preferably daily, and the grease buried or incinerated. Provide cleaning staff with appropriate personal protective equipment (PPE) and cleaning tools. The tank should be large enough to capture oils and grease but not so large that suspended solids settle out on the tank bottom. A tank liquid volume equal to the maximum hourly sullage flow rate is suggested (Medicins Sans Frontieres, 2010). The top of the trap should be at least 100 mm above ground level and covered to prevent the entry of surface water and to prevent accidents. Traps can be built from bricks, blocks, plastic, wood, or an oil drum cut in half along its longest axis. Box 14.2 provides an outline design example.

Box 14.2 Design of a grease trap

A grease trap is required to remove grease, oil, and gross solids from a block of four bathing cubicles, each fitted with a tap and a shower.
Based on Table 14.2 a wastewater flow from each user is 100 litres
Assume a maximum of two users per cubicle per hour
Therefore, the peak hourly flow = $100 \times 2 \times 4 = 800$ litres/hour
Construct a square tank (similar to a septic tank)
Assume a liquid depth of 1.0 m
Surface area is 0.8 m²
Assume a length to breadth ratio of 2:1
Width = $(0.8 \div 2)^{1/2} = 0.63$ m
Tank length = 1.3 m
Divide the tank into two compartments where the first compartment is around 70 per cent of the total volume, as shown in Figure 14.2.

Advantages
- Effective at removing gross solids, oils, and fats
- Relatively simple to design and construct

Disadvantages
- Requires frequent cleaning to remove gross solids, oils, and fats
- Must be located close to the sullage source

14.5.3 Surface application

Where sullage quantities are small, such as households using water carried from a communal water point, simply throwing the sullage on the ground near the home may be satisfactory. Warm, dry climates with permeable soils are well suited to this method but it can create local environmental issues in cold, wet climates with impermeable soils. The main issues are standing water and a build-up of organic material such as food waste; this attracts flies and creates odours. In practice, this is the most likely practice for most residents if left to their own devices. Observation will quickly show if the practice is unsuitable and alternative disposal methods are required.

14.5.4 Settlement tank

For larger flows, such as from a hospital complex, a grease trap will be insufficient and a settlement tank is required. Similar to a small septic tank, a settlement tank (Figure 14.3) collects suspended solids, detergents, and oils prior to disposal. The settled solids and scum should be removed and buried when they take up about one-third of the liquid volume. Table 14.3 suggests typical tank sizes for different sullage flows. Smaller tanks are frequently constructed from vertical concrete rings: one for the primary settlement and a second for the secondary settlement (Figure 14.4).

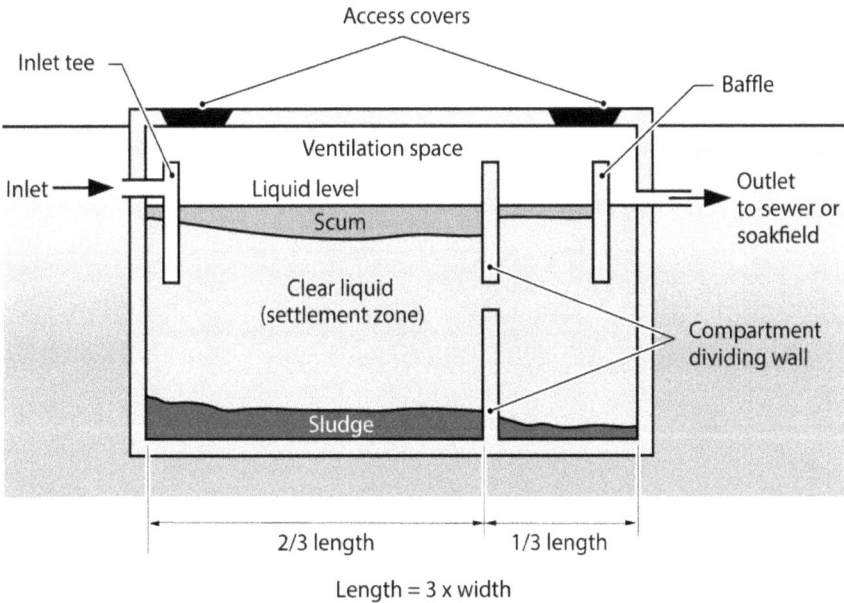

Figure 14.3 Rectangular sullage settlement tank

Table 14.3 Sullage settlement tank sizes

Inflow rate (litres/day)	Rectangular tanks Liquid depth^ (m)	Tank length^^ (m)	Tank width (m)	Circular tanks 1st tank volume (litres)	2nd tank volume (litres)^^^
2,000	1.2	1.9	1.0	1,520	760
5,000	1.4	2.8	1.4	3,660	1,830
10,000	1.5	3.3	1.7	5,610	2,810
15,000	1.5	3.4	1.7	5,780	2,890
20,000	1.5	4.0	2.0	8,000	4,000

Note: ^Allow 30 cm extra tank depth above the liquid level for ventilation.

^^The length of the first compartment is twice the length of the second.

^^^If the secondary tank volume is too large, the flow could be split between multiple tanks.

Source: Adapted from (** Harvey, et al., 2002).

The ease and speed of construction outweigh the less efficient use of space. In large institutions such as hospitals and medical centres, septic tanks can be used for the disposal of excreta and sullage combined (see Chapter 9, section 9 for more details).

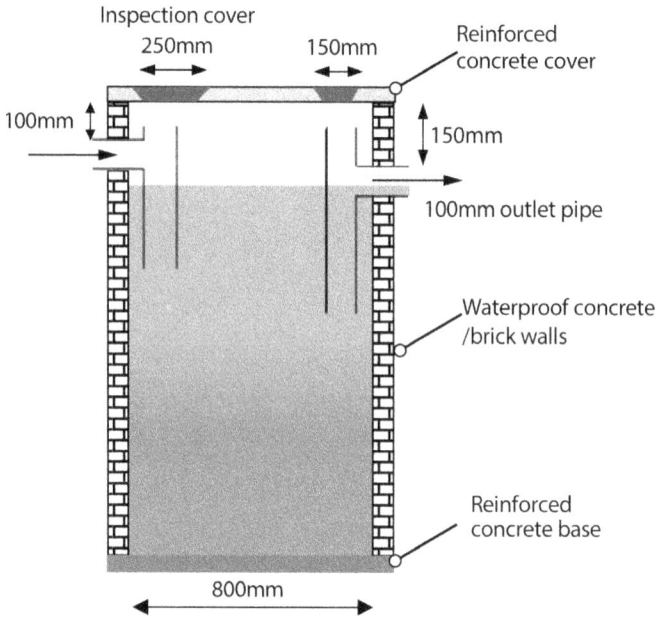

Figure 14.4 Circular sullage settlement tank

Advantages
- Cost-effective method of preparing sullage for further treatment or final subsurface disposal.
- Minimal maintenance requirements.

Disadvantages
- Grease traps may also be required if the settlement tank is a long distance from the sullage source.

14.5.4 Constructed wetlands

Sometimes known as reed beds, constructed wetlands can be used for removing organic matter, oxidizing ammonia, reducing nitrate concentrations, and removing phosphorous from sullage. Both horizontal and vertical flow beds can be used depending on the locality. Further details on their design and construction are given in Chapter 9, section 14. Failure to remove soaps and oils from the sullage prior to application will lead to rapid clogging of the filter medium.

14.6 Disposal technology options

Whenever possible, sullage should be disposed of close to the point of origin. The most common method in emergencies is probably discharge into surface water. However, there is a range of options.

Infiltration

It is common to infiltrate sullage into the surrounding soils using soak pits or infiltration trenches. Details of the design and construction of these is given in Chapter 10, section 1. Infiltration systems are easily blocked by oil, grease, soap, and suspended solids so these must all be removed from the sullage prior to infiltration. Suggested infiltration rates for sullage are 50 per cent of those given in Table 4.3.

Evapotranspiration

Evapotranspiration (ET) is the combining of moisture with the atmosphere. There are two principal methods: evaporation ponds and ET beds.

Evaporation ponds are appropriate only for small quantities of sullage (see Chapter 10, section 3 for more details). Organic material, oils, and fats must be removed prior to using this technology.

ET beds rely on capillary action to draw water to the surface of shallow sand beds for evaporation to the atmosphere. ET beds are planted with grass or other vegetation; this increases the movement of water from the root zone for transpiration through the leaves. There is very little storage for any rain that falls on a bed and this must also be removed by evaporation, so beds are suited only to dry, arid climates. Figure 14.5 shows the cross-section of one type of ET bed. The performance of ET beds depends on a range of factors including climate, hydraulic loading, capillary rise, soil, vegetation, and construction. Oils and soaps must be removed from the sullage before discharging to ET beds.

Design

ET rates vary widely (0.2–20 litres/m^2/day) depending on local conditions, but a well-designed bed in a suitable climate can dispose of around 10 litres/m^2/day.

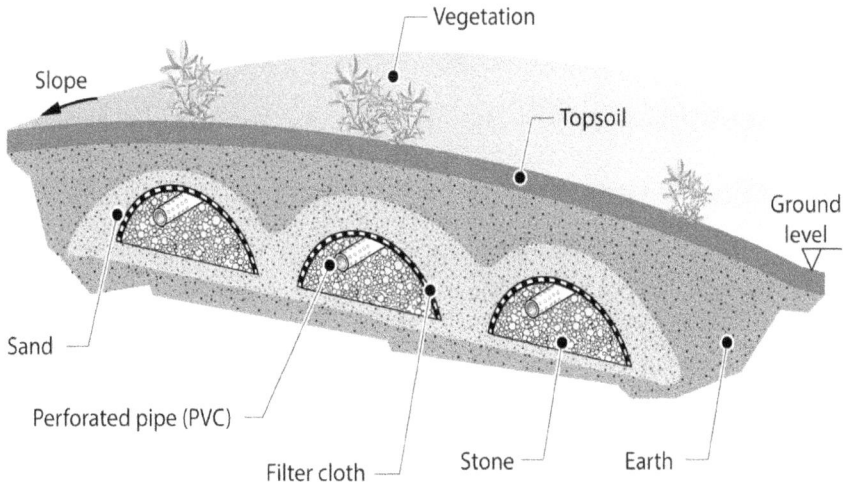

Figure 14.5 Evapotranspiration bed

Generally, the required bed surface area is divided between two beds used alternately to prevent overloading and to allow time for the soil to re-oxygenate. A layer of sand is placed in the bottom of the bed followed by a layer of gravel of a uniform size, ranging from 20 mm to 50 mm diameter, filling the bed to a depth of 300 mm or less, depending on the bed's overall depth. Distribution pipes are bedded in the gravel no more than 1,220 mm apart and no more than 610 mm from the bed walls. The top of the distribution pipe must be flush with the top of the gravel media. A water-permeable soil barrier (e.g. a geotextile filter fabric) is placed over the gravel and fine sand or loam soil added to fill the bed to within 50 mm of the top. Selecting the proper soil is extremely important in building an ET system as fine sand or loamy soils draw water towards the surface faster than coarse sand. Wicks (a column of soil that extends from the topsoil through the gravel media to the bottom of the bed) incorporated into the gravel media draw water continuously from the gravel into the soil and towards the surface area, where it evaporates or is taken up by plants. The total wick area should be 10–15 per cent of the bed surface and should be uniformly spaced throughout the bed. Above the loamy soil, the final 50 mm is filled with sandy loam and mounded in the centre with a slope of 2–4 per cent towards the outside of the bed. The last step is to plant vegetation specially selected to transpire the most water, such as certain species of grass. Placing clumps of ready grown grass over the bed may be the best approach to establishing grass there. Using seed may let the mounded soil wash away during heavy rainfall before the grass is established. Larger plants with shallow root systems, such as evergreen bushes, may also be used to help take up water. It is important that the plant mix used ensures a full year growing period.

ET beds are complex systems requiring careful design and are not recommended for the acute phase of an emergency. Further information is available from the (US Environmental Protection Agency, 2000).

14.6.1 Irrigation

Sullage with a low organic content is highly suited to irrigation. The best plants for irrigation are fast-growing fruit and vegetables or slower-growing trees and shrubs. The key consideration is the matching of sullage generation volumes and times with the demand from the crops. Oils, grease, and soap must be removed from the sullage before irrigation to prevent soil blocking and salinization. Also, sullage may contain faecal pathogens and therefore must not be used for direct application on salad crops or soft fruits such as strawberries (WHO, 2013). Most crops do not grow all year round and require water only during the growing season, so matching sullage flow to crop demand can be very difficult and requires careful monitoring. Figure 14.6 shows a small-scale pipe-based irrigation system using the wastewater from a tap stand. Note the silt and grease trap after the tap stand apron; this protects the irrigation pipes from becoming blocked. In the acute phase of an emergency, the pipe network could be replaced by simple earth channels.

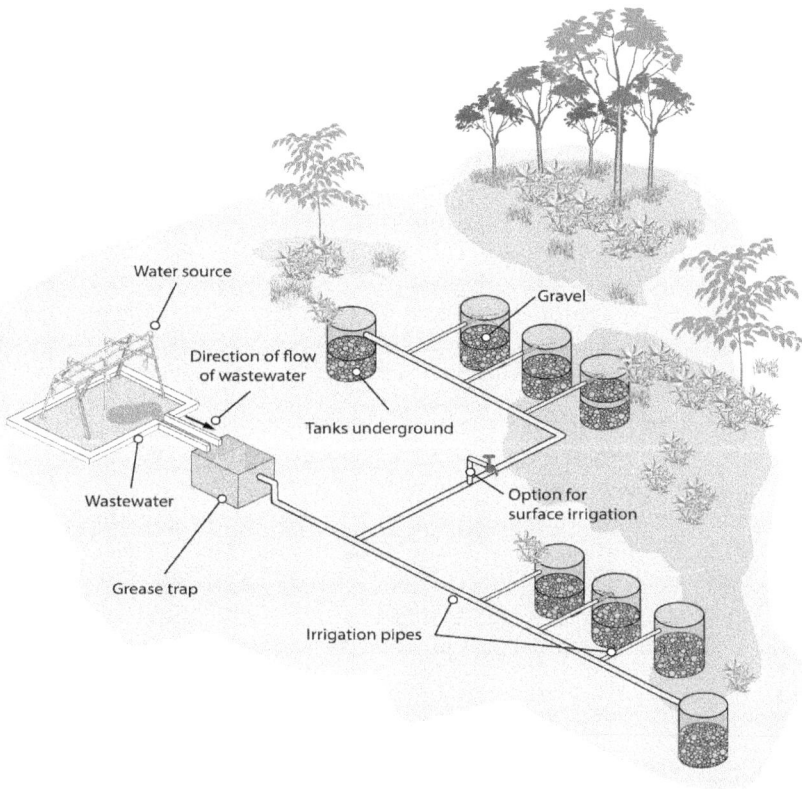

Figure 14.6 Tap stand sullage disposal via irrigation pipes

As a short-term measure, irrigation can be a highly suitable option, especially where displaced people are encouraged to grow their own crops as food supplements. Garden plots can be linked to nearby sullage generation points (such as tap stands) to provide a suitable water source. In the longer term, holding ponds can be constructed to retain sullage outside the growing season, although there are significant health and safety issues with this approach. Such projects require careful management to prevent garden users deliberately 'wasting' tap water to irrigate their crops. More information on irrigation can be found in (Erlenbach, 1998) and (Adam-Bradford, 2017)

14.6.2 Surface water diffusion

Although the disposal of untreated sullage into natural and man-made water courses is very common, it cannot normally be recommended. The high organic content of most untreated sullage will be a major pollution issue affecting wildlife, flora, and possibly downstream water users. If disposal to surface water is the only option, then:

- pre-treat the sullage to remove oil, soap, grease, and suspended solids;

- ensure a high level of dilution of the sullage with the surface water;
- discharge upstream of an area of turbulence in the surface water to ensure rapid mixing and oxygenation.

More information on surface water discharges is given in Chapter 10, section 5. Where discharging into a dry drainage ditch is the only option:

- pre-treat the sullage;
- remove obstructions from the ditch to provide an unhindered flow;
- fence off the ditch to keep out children and animals;
- observe how far the sullage travels before it is either infiltrated into the ground, evaporated, or cleaned by natural processes.

In some cases, especially in urban areas, sullage may have to be discharged to stormwater drains. This is not ideal as it contaminates stormwater, but it may be the only option during the early stages of an emergency response. Adding sullage can cause problems, particularly during dry periods when there is no stormwater flow. Small sullage flows in a channel designed to carry large stormwater flows often become contaminated with organic materials carried in the sullage; this leads to offensive smells and fly breeding and attracts vermin. The situation can be improved by creating a smaller channel in the bottom of the main channel to carry sullage during dry seasons (Figure 14.7).

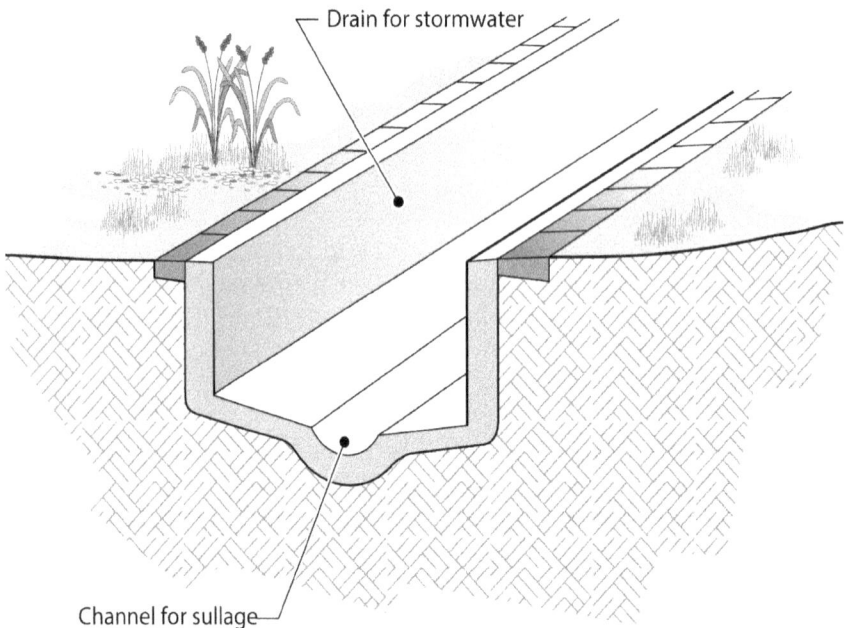

Figure 14.7 Stormwater channel incorporating a sullage channel

14.6.3 Local reuse

In emergencies it is difficult to reuse sullage because of the difficulty of ensuring its proper treatment prior to reuse. Also, in areas where water is limited, users may be tempted to use partially treated sullage as a potable source. However, there are some tasks for which treated or partially treated sullage may be appropriate, such as the ones listed in Box 14.3. Extreme care must be taken to keep sullage water for reuse completely separate from potable water sources. Separate containers should be retained for sullage storage and for potable water storage.

Box 14.3 Possible uses for partially treated sullage in an emergency

- flushing toilets;
- washing floors;
- cleaning vehicles;
- watering small gardens;
- brick and block making;
- drinking water for livestock and pack animals;
- dust suppression.

14.6.4 Sullage disposal in very cold climates

The disposal of sullage in very cold climates is problematic, particularly in high-density communities. Sullage generally has no value and therefore it is difficult to get communities to take its disposal seriously. However, failure to dispose of it properly can lead to serious hazards such as surface icing. Supporting or building on current practices is usually the best solution, as those practices are already accepted and understood by the users. Where that is not possible, consider one of the following:

- **Minimize and concentrate sullage volumes:** Tight management of water use and engagement with the affected community on the best to ways to minimize sullage generation. Alternatively, activities that generate sullage such as bathing, laundry, and food preparation can be centralized in a communal building that is more economical to heat and can provide a common collection point for sullage.
- **Storage:** Storing sullage in underground tanks is possible for single households, centralized services, or small institutions. However, costs become excessive for larger population sizes. All discharge pipes and access chambers, and sometimes even the storage tank, must be highly insulated to prevent freezing.
- **Infiltration below the frozen zone:** Generally, the ground freezes to only a shallow depth. Constructing an infiltration system such as soakaway or infiltration trenches below that depth may be possible

if the soil profile is appropriate (see Chapter 4, section 4). Sometimes soakaways can be constructed partially in the freezing zone. The volume in the frozen zone provides additional storage for sullage during the cold season which can percolate into the soil during warmer periods. All pipes above ground and running through the frozen soil zone must be either insulated or heated to prevent freezing (see Chapter 11 for more details).

CHAPTER 15
Emergency solid waste management

Poor solid waste management (SWM) has many negative consequences for communities. It can adversely affect general well-being, create a health hazard, and impact on the wider environment. During an emergency the disposal of solid waste or rubbish can become a critical issue as existing disposal and collection systems are likely to be inoperable or heavily disrupted. On new sites such as refugee camps, there will be no waste management system in place, requiring immediate planning in order to implement one. Crisis situations often cause extra waste, such as flood debris or rubble from destroyed buildings. This must be dealt with quickly as it can hinder emergency relief efforts, demoralize the affected community, and discourage efforts to improve other aspects of environmental health. Humanitarian responses also generate waste, such as emergency supplies packaging and medical waste. This is a significant issue throughout an intervention, from first response to the decommissioning of sites.

This chapter covers both general and healthcare solid waste. In many ways they are very different, but there is some overlap. Much of the background sections are applicable to both waste streams but the sections related to storage and disposal are significantly different. Specific details of healthcare waste are covered in Chapter 15, section 13.

In this book, the term 'solid waste' is used to include all primarily non-liquid wastes generated by human activity and those resulting from the disaster itself including those listed in Box 15.1

Box 15.1 Sources of solid waste during and following a disaster

- general domestic, institutional, and commercial solid waste as listed in Box 15.2;
- human faeces disposed of in solid waste;
- commercial food processing and packaging waste;
- waste from the emergency response such as plastic water bottles and packaging from other emergency supplies;
- building rubble resulting from natural disasters and military interventions;
- mud and slurry deposited by natural events;
- fallen trees and rocks obstructing transport and communications;
- medical waste from hospitals, clinics, etc.;
- general and toxic waste from industry.

Box 15.2 Classical waste streams

Organic matter/food waste	Lead-acid batteries
Paper/cardboard	Household batteries
Glass	Used engine oils
Metals	Paints, solvents, and varnishes
Plastic	Electronic waste
Rubber	Medical waste
Miscellaneous combustibles	Other toxic wastes
Miscellaneous incombustibles	

15.1 Hazards of poor solid waste management

The hazards from poor SWM are numerous and varied. Table 15.1 identifies the most common problems associated with poor SWM while Table 15.2 lists the specific issues related to individual disaster types.

Table 15.1 General hazards from unmanaged solid waste

Health	Environmental
• Scavenging for food leading to increased cases of disease (e.g. dysentery).	• Toxic household wastes (in particular domestic batteries, vehicle batteries, used vehicle oils, paints, solvents, varnishes, aerosols, pesticides, fluorescent lightbulbs, asbestos, and broken electrical apparatus) can contaminate the drinking water or food chains and can be toxic or fatal to human health, plant, animal, and aquatic life or ecosystems, even in small quantities.
• Drainage networks blocked with uncollected waste leading to the breeding of mosquitoes that transmit malaria, dengue, and yellow fever.	
• Fire and smoke from heaps of solid waste may create a health hazard (i.e. burning plastics or chemicals).	
• Aerosols from dust and fungi in uncontrolled waste dumps can cause respiratory health problems.	• Solid waste washed by rain can create strong leachates that contaminate ground and surface water, leading to negative environmental effects on ecosystems.
• Medical and military waste present health and physical hazards to people walking through the area.	
• Flies, rats, dogs, snakes, and other scavengers create an increase in disease vector populations and the spread of infectious diseases.	• Indiscriminate dumping of waste can block water courses and cause flooding.
• Waste is unsightly and lowers the morale of communities.	• Inappropriately managed inert non-biodegradable wastes (e.g. plastics and micro-plastics) can cause long-term damage to animal life, especially if ingested. A bigger potential problem is micro-plastics.

Table 15.2 Emergency types and their waste characteristics

Emergency	Hazards
Earthquakes	• Structures collapse 'in-situ', trapping waste within damaged buildings and structures and creating challenges in sorting hazardous materials from non-hazardous
	• Site access frequently blocked by collapsed buildings
	• Quantities of waste are high compared with other disaster types since all building contents normally become waste
	• Flooding from water pipes and sewers leads to widespread faecal contamination
Flooding	• Mass displacement of buildings, infrastructure, and earth, generating large volumes of building materials, household waste, mud, gravel, and sometimes mixed hazardous materials
	• Access roads blocked by damaged roads, bridges, and flood-washed materials
Tsunami	• Widespread damage to infrastructure with mixed debris spread over a large area
	• Ships and boats thrown ashore or sunk, requiring specialized waste management skills
Hurricanes, typhoons, cyclones	• Widespread damage to roofs and some buildings may collapse
	• Damage to ships and boats
	• Damage to above-ground utilities with possible contamination from hazardous wastes (e.g. transformer oil)
Conflict	• Widespread damage to buildings and infrastructure from military hardware
	• Local waste collection infrastructure damaged or stolen
	• Unexploded ordinance
	• Damaged sanitary facilities leading to environmental faecal contamination

Source: Adapted from (** UNEP/OCHA Envionment Unit, 2011)

15.2 Responsibility for service delivery

Responsibility for emergency SWM is sometimes unclear. In theory, as with all public utilities, ultimate responsibility lies with the host government, although it is often devolved to local authorities, which, in turn, may subcontract some or all of waste management to private contractors. In emergencies, where no international relief agencies are involved, this is most likely the case. Where support from international agencies is necessary, either because the host government cannot cope or because there is no host government,

Table 15.3 Responsibility for SWM provision in emergencies

Agency	Coordination unit	Responsibilities
UNDP^	Early recovery cluster	Management of waste generated by the disaster
UNICEF^^	WASH Cluster	Solid waste generated by the affected population after the disaster outside specific environments (see below)
		Provides advice and support to specific environments
WHO^^^	Health Cluster	Medical waste management within health facilities
UNICEF^^^^	Education Cluster	Solid waste management within education facilities
UNHCR		All SWM in protracted crises

Source: ^ (UNDP Crisis Prevention and Recovery, n.d.)
    ^^ (Global WASH Cluster, 2009)
    ^^^ (World Health Organization, 2020)
    ^^^^ (Global Education Cluster, 2010)

then responsibility is generally divided between a number of UN agencies, as shown in Table 15.3. In most cases, the work is devolved to other agencies (and sometimes contractors), usually in support of the established national waste management organizations.

15.3 Importance of expert advice

The implementation of solid waste collection, treatment, disposal, and recycling programmes in emergency contexts can be complicated by additional constraints. These include:

1. sites that are difficult to access with waste collection vehicles;
2. lack of adequate final waste disposal sites;
3. lack of adequate and safe waste collection and processing equipment;
4. unfamiliarity or unwillingness of the population to use waste collection containers, or to reduce, recycle, or reuse waste;
5. complex environmental or public health legislation controlling waste activities;
6. complex urban environments;
7. reliance on existing waste facilities that have been poorly managed for decades;
8. loss of waste collection and processing staff;
9. budget limitations, especially in protracted crises.

In large emergencies, it is essential to seek expert advice from professionals who are familiar with the context. Solid waste vehicles, waste containers, processing equipment, and treatment and disposal technology that work well in one context may completely fail in another. In all cases, the approaches used must build on local practice, existing supply chains for processing waste, and the expertise of local specialists. Assistance should be sought locally from sources such as government departments, the UN system, NGOs, universities, consultants, or contractors who have specialist local knowledge and experience of SWM.

15.4   Objectives of emergency solid waste management

There is no comprehensive objective for an emergency SWM intervention; the nearest is that used in SPHERE, as shown in Box 15.3.

Box 15.3 SPHERE standards for SWM

SPHERE has three standards related to SWM but only the first relates to SWM as a whole. The other two relate to specific elements of SWM.

SPHERE standard one states that:

Solid waste is safely contained to avoid pollution of the natural, living, learning, working, and communal environments.

Indicator of success:

There is no solid waste accumulating around designated neighbourhood or communal public collection points.

Key actions:

1. Design the solid waste disposal programme based on public health risks, assessment of waste generated by households and institutions, and existing practice.
   • Assess capacities for local reuse, repurposing, recycling, or composting.
   • Understand the roles of women, men, girls, and boys in SWM to avoid creating additional protection risks.
2. Work with local or municipal authorities and service providers to make sure existing systems and infrastructure are not overloaded, particularly in urban areas.
   • Ensure that new and existing off-site treatment and disposal facilities can be used by everyone.
   • Establish a timeline for complying as quickly as possible with local health standards or policies on SWM.
3. Organize periodic or targeted solid waste clean-up campaigns with the necessary infrastructure in place to support the campaign.
4. Provide protective clothing for and immunize people who collect and dispose of solid waste and those involved in reuse or repurposing.
5. Ensure that treatment sites are appropriately, adequately, and safely managed.

   • Use any safe and appropriate treatment and disposal methods, including burying, managed landfill, and incineration.
   • Manage waste management sites to prevent or minimize protection risks, especially for children.
6. Minimize packing material and reduce the solid waste burden by working with organizations responsible for food and household item distribution.

Source: (** Sphere Association, 2018)

15.5 Action planning

The most important actions in an emergency change with the various phases of the response. This chapter describes action planning where international agencies are involved but a purely local response would be very similar.

15.5.1 Acute phase

Immediately after the disaster occurs, the primary objective is to tackle waste issues required to save lives, alleviate suffering, and facilitate rescue operations. Any other considerations at this stage are secondary.

Immediate actions
These are often initiated by the affected population within hours of the disaster event and quickly supported by national emergency response agencies such as the armed forces and civil defence agencies.

Create a hazard ranking using the following steps to identify the most urgent priorities:

- **Identify waste,** including potential highly toxic waste or unexploded weaponry that may need specialist handling. Source information from a site investigation, governmental sources, geographical information system, news, and information gathered from the local population and agencies.
- **Characterize waste.** Map and quantify the composition and quality of the identified waste streams (see Chapter 15 section 6). Assess existing garbage dumps and landfill sites through site visits and discussions with key stakeholders. In the early stages, waste sampling or analysis will be cursory but is still worth doing. The map will be a valuable tool throughout the process and can be updated as information becomes available.
- **Prioritize actions**. Establish a framework of waste disposal priorities, disposal points, and transport routes. This is not a purely objective process and must be based on local conditions. Each identified waste stream and/or issue is given a 'common sense' ranking based on the following as a guide:

  - Protect critical infrastructure such as water sources, food supplies, and hospitals.
  - Identify appropriate disposal sites for the disposal of the different types of waste collected in the acute phase. If an existing disposal site is available, it should be rapidly assessed for environmental compliance before use. Where no existing disposal sites are available, temporary transfer stations, disposal sites, or engineered waste sites (see Chapter 15, section 9.3) should be identified and established.

- Main streets are to be cleared to provide access for search and rescue efforts and relief provisions. Any disaster waste moved should, if possible, stay in the emergency area. It should not be moved out before appropriate disposal site(s) have been identified.
- All available equipment and stakeholders should be used. Wheelbarrows and wooden carts pulled by animals can be used where excavators, trucks, and skips do not have access.
- If hospital and clinics are affected by the disaster, they should be encouraged to separate infectious and/or healthcare waste, store it separately, and transport it to temporary special treatment or disposal sites (see Chapter 15, section 13).
- Mobilize affected and nearby communities to assist in the collection and removal of solid waste. Provide them with the appropriate tools and personal protective equipment (PPE) (see Box 15.4 for more information). Pay them if possible as this will have wider benefits for the community.

Following actions

These are often initiated within days of the disaster event:

- If people remain in the disaster area, the collection of their household waste should be reinstated where possible.
- At this stage a rapid disaster waste assessment (see previous section) should be carried out to inform further decision making. Exact data is not required, but reasonable ideas about the status of waste, the ability of local authorities to handle the situation, and the need for any national or international assistance should be provided.
- Wastes from temporary settlements should be managed in coordination with general SWM services, and thus integrated with the local waste collection services.

Box 15.4 Recommended tools for a clean-up brigade of 20 persons

Quantity	Item
10	Shovels
5	Hoes
5	Rakes
5	Wheelbarrows
20	Gloves (pairs)
20	Boots (pairs)
200	Face masks (at least one fresh mask a day)
20	Safety helmets
20	Overalls
5	Backpack sprayers
20 litres	Bleach (5–7%)

Quantities are for guidance only and should be adapted to context.

- Ownership of waste, in particular reusable waste, is an important issue to clarify to avoid later conflicts.
- Establish relationships with the affected communities and how they would like to be consulted. Determine their attitudes and priorities related to SWM.

The output from these actions should be an understanding of the basic issues and a series of actions to address the most urgent of these.

15.5.2 Stabilization phase

The stabilization phase lays the groundwork for a disaster waste management programme to be implemented during the recovery phase. It also continues to address key issues such as the location of disposal sites for the different types of waste, streamlining logistics for waste collection, transportation, and reuse/recycling activities. Efforts build on the acute phase assessment but go into greater depth with an emphasis on longer-term solutions. Required actions normally include the following.

Assessments
- Continue to assess the wastes produced by the disaster (quantities, locations, types of waste, regulatory understanding, etc.) (see Chapter 15, section 6).
- Assess separate locations for medium-term waste separation and temporary disposal sites for disaster-generated rubble and municipal waste (and keep them separate). This may entail upgrading or improving current disposal sites and recycling facilities.
- Assess requirements to close current waste disposal sites if they pose a threat to human health or the environment.
- Identify and assess other waste management facilities in or near the disaster affected area.
- Assess local capacities for addressing emergency waste management and identify gaps or needs for additional assistance.

Planning
- Develop a plan for healthcare waste. This may entail the construction of a temporary incinerator for healthcare waste (see Chapter 15, section 13).
- In collaboration with specialist providers, develop a plan for hazardous and toxic waste collection, treatment, and disposal.
- Waste containers, collection vehicles, and transport services should be planned together, as they depend on each other.
- Identify exit strategies and handover options for the disaster waste management systems planned for or established.
- Develop an outline budget for the intervention and submit it to the authority in charge and to possible donors.

Operations
- Implement or strengthen a communications plan for the affected communities with a focus on opportunities (i.e. reuse and recycling), risks (i.e. human health risks), and collection schemes. Also discuss issues relating to public health, livelihoods, and the environment.
- Establish temporary storage sites for debris and regular waste. (Debris can often be reused as it is often relatively inert.) It is better to keep it stored separately for processing at a later date than for it to clog up the regular solid waste infrastructure.
- Initiate the collection and transportation of waste and debris, with the goal of expanding this in the full recovery phase.
- Prepare practical advice and guidance for local authorities on interim solutions to minimize the environmental and health impacts of disaster-generated waste.

Communication and reporting
- Communicate rapidly and regularly regarding all findings to the local authorities, and to other national and international response organizations as appropriate.
- Document the assessment results, recommendations, and mitigation measures implemented. The outputs from the planning actions include data and information to design the disaster waste management programme to be implemented in the recovery phase.

15.5.3 Recovery phase

This phase includes implementation of waste management projects designed in the stabilization phase and revised through continued monitoring and evaluation of the situation. The following main actions should be considered:

- Update the programme budget and secure long-term funding.
- Update and implement a communications plan for the key stakeholders to ensure that the waste management programme is aligned with community expectations and needs and with any existing waste management system.
- Procure or repair waste management plant, machinery, and equipment.
- Train waste management operators, if required.
- Support the implementation of the waste management plan by strengthening waste management operators/operatives or local authorities through the provision of additional funds, training, or specialist staff.
- Combine the disaster waste management service with existing waste management services ready for formal handover.

The output of this phase should be a fully functioning SWM service serving the disaster-affected communities and tailored for merging with existing SWM services.

Table 15.4 The waste hierarchy

Hierarchy	Intervention	Description
	PREVENT	Preventing or reducing the amount of waste generated by reducing the packaging. Designing products for longer life and using less hazardous materials, etc.
	REDUCE	Reducing the toxicity or negative environmental or health impacts of the waste.
	REUSE	Reusing, repairing, and reselling used items. Includes direct reuse of items in their original form or function, e.g. bottles, plastic jars, and boxes.
	RECYCLE	Turning wastes into raw materials for new products, e.g. glass, metals, and thermosetting plastics. Includes composting of organic wastes.
	RECOVER	Recovering energy in certain types of waste by incineration, anaerobic digestion, gasification, or pyrolysis or similar processes.
	DISPOSE	Disposing of wastes that cannot be reused, recycled, or recovered in an environmentally sound manner.

Source: Adapted from (** Harvey, 2020)

15.6   Waste hierarchy and recycling

The waste hierarchy is a classification of waste management options in order of their public health and environmental importance (Table 15.4). It is very unlikely that formal waste reuse and recycling initiatives can be introduced during the acute phase of an emergency, but they could be gradually introduced during the stabilization phase. The waste hierarchy highlights those interventions that reduce the toxicity or negative impacts of waste that must be prioritized over interventions that lessen the quantity of waste or affect its reuse, recycling, or disposal. An example of this in an emergency setting is the prioritization of medical or toxic special wastes of which small quantities can have a major impact on public health and the environment.

15.6.1 Waste reuse

Waste reuse involves repairing, refurbishing, reselling, or reusing items in their original form or function (e.g. bottles, plastic jars, or boxes). In emergency settings there is a significant market for the repair of broken goods or the sale of used goods. These activities should be encouraged, strengthened, or reinvigorated. Not all 'waste' is waste: metals, glass, and plastic, in addition to any materials that can be burned, are usually scavenged very quickly. Recycling and reuse are also very common at the household level, particularly of any materials that can be burned, fed to animals, or turned into something else. Due to the social and economic nature of recycling, it should always be organized in close collaboration

Table 15.5 Recyclable (salvage) materials

Material	Comments
Glass	Easy to collect, suitable for creating new objects, or reuse for another purpose.
	Possible buyback by product manufacturer.
Metals	Aluminium has high bulk purchase price.
	Metal containers can be sold on or reused/converted to other products.
Paper	Bulk forward selling, especially if graded.
	Can be used as a heat source, as insulation, and for reuse.
Plastics	Thermoplastic (e.g. PVC) can be remoulded to new shapes.
	Bulk forward selling is possible if clean, good quality, and sorted by colour.
	Thermoset plastics (e.g. appliance casings) are not recyclable.
	Recycling is complex because of the wide variety of plastics manufactured and their different recycling opportunities (some cannot be recycled).
Rubber	Can be recycled as hard-wearing products such as shoes, belts, or small containers.
Clothing, bedding, etc.	Can be reused as is or converted to other products, such as clothing, insulation, blankets, etc.
Organic waste	Can be converted to fertilizer through composting or used for biogas generation.

with representatives of the affected population and preferably undertaken by the affected population. Table 15.5 lists the most common materials suitable for recycling. Waste sorting at the beginning of the collection process is easier and produces cleaner products than sorting at the disposal point. However, it is difficult to persuade poor communities to participate in recycling unless they can see some tangible benefit; it may be necessary to offer some incentive to encourage them to join in. Some form of processing of recyclables may be necessary prior to transportation or selling on. This may include the following activities:

- crushing for aluminium, steel cans, etc.;
- shredding of papers, cardboard, tyres, etc.;
- bailing of fibrous materials such as fabrics, cardboard, etc.

Depending on the facilities and technologies available, all or some of these processing options will be viable. The processing of recyclables increases the density of the materials involved, making them more efficient and economical to transport and resell (Oxfam, 2008).

15.6.2 Manual waste sorting

Manual sorting of wastes at a centralized level can be reasonably successful provided that some degree of source separation has been done at the household and communal level. At a most basic level, this requires the affected population to be informed and motivated to separate their wastes into 'wet' biodegradable mixed waste that goes straight to landfill and 'dry' non-putrescible waste that may be sorted later into paper, rags, plastics, glass, metals, etc. Manual sorting requires, at a minimum, a clean yard area in which the waste can be tipped and bins into which the materials can be manually sorted. Improved setups use a sorting table with large corner holes into which wastes can be directed into bins without bending over. Identify potentially dangerous materials such as dangerous liquids, gas, inflammable materials, batteries, or poisonous objects that should be handled with special care. Provide separate containers for these dangerous materials; they should be kept in a specific area of the site.

Waste sorters must have good access to handwashing and showering facilities, be provided with suitable PPE, and be trained in infection control standard precautions. Waste handling and sorting frequently involve moving loads of waste; this should be facilitated with handcarts and wheelbarrows. A hoist is invaluable if bales of recycled materials need to be loaded onto a vehicle. A set of scales is also essential to document the amount of recycled material that is produced. Finally, fire extinguishers, buckets with sand, and alarm bells placed at several visible locations are essential in case of fire. Table 15.6 gives approximate manual sorting rates for different materials.

15.6.3 Mechanized waste sorting

Unless a mechanized waste sorting facility already exists, it is very unlikely that one will be built during the acute or stabilization phase of an emergency response.

15.6.4 Working with waste scavengers

It is estimated that more than 2 per cent of the population in developing countries earns their living from scavenging and it can be very difficult to prevent. Working with scavengers in a way that complements ongoing waste collection is preferable to outlawing it. Not only does it recover valuable materials, but it also creates jobs in a situation where they are in short supply. In general, any scavenging activity that removes recoverable materials from the waste stream reduces the waste disposal costs and therefore scavengers should be recognized for their contribution. One way of working is to define where and how scavenging can take place safely so that the public health risk is minimized and it doesn't disrupt waste collection or create additional mess. Scavengers could be given exclusive rights to waste collection points, provided that the allocated scavengers fulfil the PPE precautions and keep the site clean and tidy. Alternatively, they could be allowed to recover wastes in a controlled environment, such as a combined transfer station/resources recovery facility. In all cases, the scavenging of healthcare wastes must be forbidden.

Table 15.6 Manual waste sorting rates and efficiencies

Material	Sorting rate (kg/hr/person)	Recovery efficiency (%)
Paper	700–4,500	60–95
Cardboard	700–4,500	60–95
Glass	400–800	70–95
Plastic	140–280	80–95
Aluminium	45–55	80–95

Source: (** Harvey, 2020)

15.7   Measuring solid waste characteristics

The most accurate method of estimating waste characteristics is by using a sampling programme. Since it is not feasible to analyse large quantities of waste, the samples that are used must represent the whole mass of waste.

Before conducting a sampling programme, it is useful to ask the following questions:

- Why are you conducting this sampling? In the early stages of an emergency response, the main purpose is to assess the approximate quantities and composition. As the response progresses, a more accurate assessment is necessary to check for changes in waste content and quantity and to select a medium-term disposal strategy.
- Do you have any secondary data or information that can be used to crosscheck results? In urban emergencies, existing solid waste disposal organizations should have data on waste characteristics, but this must be revised to allow for infrastructure damage and changes in population numbers, density, and waste disposal methods.
- Do you understand the present system? This is particularly relevant for how different types of wastes are added and separated.
- Have you chosen appropriate times for sampling to represent the various days of the week and the seasons? It is vital that these factors are considered when taking samples and using results. The seasonal availability of fruit and vegetables and the need for heating in cold weather are two issues that cause seasonal variations.
- Have you chosen the appropriate stage of the waste system for sampling in order to obtain the information desired?
  Normally, you should select the stage immediately before your organization becomes involved.
- What type or area are you sampling (residential, market, non-food items (NFI) distribution, health centres, etc.)?
  Areas with a marked difference in waste generated should be sampled separately.
- Are you selecting the appropriate sample size, keeping in view the required degree of accuracy, budget, organizational capacity, etc.?

15.7.1 Sample size

The larger the sample, the more representative it will be of the waste generated but the more difficult it will be to assess. It is usual to divide the waste collection area into sub-areas and take random samples from waste storage points within each area. Always collect all of the waste from each sampling point and ensure that all the sampling points within a sub-area have waste collected over the same number of days.

At the start of a programme, four to eight randomly selected samples are sufficient. Remove all wastes from the sampling points prior to beginning sampling so that only wastes deposited during the sampling period (typically three or four days) are measured. A rapid census is also required at each collection point to understand how many people are generating the waste.

15.7.2 The importance of the point of sampling

In the acute phase, when waste streaming is not possible, waste sampling can take place at any point in the waste management process. If waste is deposited in shared (communal) containers in the street, it may be necessary to provide temporary household containers to better understand the waste generation of individual families. An estimate of the number of families using each container is sufficient. In many low-income countries there is enormous potential for change in the waste characteristics as the waste goes through the system from generation to disposal (Table 15.7). Select the stage in the system where your organization intends to begin support. A simple method for sampling the waste stream is given in Box 15.5.

15.7.3 The weight of waste

There are many ways to estimate the weight of waste being generated depending on how accurate the estimate needs to be and at what stage in the solid waste

Table 15.7 Changes in waste volume between source and disposal

Waste volume decreases	Waste volume increases
Natural biodegradation and volatilization of waste constituents	Rain soaking into the waste
Burning of the waste	Animal manure added
Eating of waste by animals	Use of storage sites as toilets or adding nightsoil
Salvage sold or used by household, servants, or street waste pickers	External waste generation sectors (such as small industries, shops, and restaurants) discarding their waste in bins intended for household waste
Scavenging by waste collectors	Silt and stone being added from street cleaning or building work
Scavenging by waste disposal site staff and waste pickers	

Source: Adapted from (Flintoff, 1976)

Box 15.5 Solid waste sampling method

1. Place the collected waste on a table whose top is made of a 50 mm steel mesh grid. Move the waste around so that all the particles smaller than 50 mm fall onto the floor below.
2. Each component remaining on the table is then picked out and placed in a box for subsequent weighing. If possible, the articles should be sorted into different boxes in line with the different disposal streams.
3. The table is frequently shaken to ensure that all undersize material falls through.
4. Place the material collected that is under 50 mm on a second table whose top is made of a 10 mm steel mesh and repeat the above process.
5. The fine matter remaining will be a mixture of inert and organic matter. The proportions of each can be found using laboratory moisture and ignition tests. With experience, however, estimations can be made visually.

An example of the results of this procedure for domestic waste are shown in Table 15.8.
   If necessary, more detailed analyses of the waste can be carried out to determine its chemical and biological properties, such as moisture and ash contents, calorific value, and concentrations of carbon, nitrogen, oxygen, and heavy metals. However, this is rarely a priority in an emergency response. These tests are generally conducted in laboratories, using small samples that are carefully prepared according to the general composition of the waste.

Table 15.8 Example waste composition analysis

Waste constituent	Weight (kg)	%
1. Paper	55.5	14
2. Metals	19.8	5
3. Glass	15.8	4
4. Textiles	11.9	3
5. Plastics and rubber	7.9	2
6. Bones	4.0	1
7. Miscellaneous combustibles	7.9	2
8. Miscellaneous incombustibles	15.8	4
9. Inert matter <10 mm^	39.6	10
10. Organic matter >50 mm^^	39.2	10
11. Organic matter 10–50 mm^^	138.7	35
12. Organic matter <10 mm^^	40.0	10
Total	396.2	100
Density (kg/m^3)^^^		387.4

Notes: ^ Sand from wind-blown dust and courtyard cleaning can be a significant proportion of the waste mass. Steps should be taken to keep this at household level.
   ^^ Organic waste is often the largest component of domestic solid waste and so deserves special attention.
   ^^^ See Chapter 15.7.4.
   The presence of additional wastes should be noted, such as light industrial wastes, batteries, used vehicle oils, paints, solvents, varnishes, aerosols, and broken electrical apparatus (including strip lights).
Source: Adapted from (** Harvey, 2020)

process the data is required. During the acute phase it is not important to fully understand the types, weights, and volumes of wastes being generated but it becomes more so when designing for the longer-term response. The volume of waste generated directly by a disaster (such as an earthquake) will vary greatly and will have to be assessed separately, although the types of waste likely to occur are shown in Table 15.2.

The weight of the waste arriving at the final disposal site will be a function of the weight of waste generated less that removed by waste pickers. One must also consider that salvage may take place at the disposal site by waste pickers who pick recyclables out of the deposited waste. The method of assessing the weight will depend on the stage in the emergency response and the equipment available for measuring. The most common methods are listed below:

- **An accurate estimate:** Weigh all the waste vehicles as they enter and leave the site. To account for changes due to seasonal or daily fluctuations, the weight survey should be conducted over a two-week period on between two and four occasions throughout the year. This method does not distinguish between various kinds of waste.
- **A less accurate estimate:** Weigh randomly selected loads ensuring that all types of truck and all types of waste are included.
- **A rough estimate:** Count the number of loads arriving per day and weigh the loads carried by a small number of randomly selected waste vehicles.

The actual total daily weight of household waste generated can be calculated by estimating the per capita rate and multiplying it by the population. Sampling must be carried out in order to find the per capita rate. The best way to do this is to set up a modest programme as described below, ensuring that all socio-economic groups are considered:

1. Provide each participating household with a container (such as a plastic bag) for the collection of each day's output of waste.
2. Collect the containers every day, ensuring that they are adequately labelled, and transport them to a central place for analysis. Ensure that the weight, number of people in the household, and area or socio-economic grouping of the area are recorded.

Samples should be collected for at least a 10-day period. The total domestic waste generation rate is then calculated by multiplying the quantities for each characteristic group by the population receiving a waste collection service in the respective area. In general, in industrialized nations the population may create up to 2 kg of waste per person per day. This waste is typically of low density, containing a high proportion of paper and plastic packaging. In less developed countries, the waste generated may be lower (see Table 15.9 for examples). In many temporary settings, waste generation is even lower – between 0.3 and 1.0 kg per person per day – and the waste is much denser, containing a greater proportion of organic matter. A similar approach can be used for waste from markets, restaurants, etc.

Table 15.9 Examples of waste creation rates and densities

Location	Rate (kg/person/day)	Density (kg/m³)
Early-stage emergency	0.5	200–400
Bangladesh (Dhaka)	0.5	
Brazil (Belo Horizonte)	1.4	
India	0.3–0.55	400–600
Indonesia	0.6	400
Mali (Bamako)	0.7	
Mexico	0.68	300–500
Myanmar		400
Nigeria	0.55–0.58	250
Pakistan	0.3–0.55	500
Sri Lanka	0.67–0.73	400
Turkey	1.2–1.1	390–520
Tunisia (Sousse)	1.1	
United Kingdom	1.5–1.3	150
United States	1.6–2.0	100
Zambia (Lusaka)	0.6	

Source: Adapted from (** Harvey, 2020), (UN-HABITAT, 2010) and (** Sphere Association, 2018)

15.7.4 Waste density

An estimate of waste density is important as it defines the volume necessary for storing the waste collected.

Method 1
1. Fill a container of known volume and weight with a sample of waste, ensuring that it is shaken gently during loading, but not compacted. (1 m³ is considered an appropriate size of container.)
2. Measure the weight of the waste.
3. Calculate the bulk density by dividing the weight of the waste (kg) by the volume of the container (m3).
4. Repeat several times in order to obtain a reliable average.

Note: The density is different at different stages of collection and may be higher at the end of a journey.

Method 2
The density of waste in open trucks can be estimated by measuring the volume of the load-carrying body and estimating the proportion of this volume that is filled with waste. Weighbridge results can be used to determine the weight

of each load. This should be repeated for several truckloads. International examples of waste density are shown in Table 15.9.

15.7.5 Moisture content

The moisture content is important for assessing the risk of leachate problems and how well organic matter is likely to decompose. It is determined by preparing smaller portions of waste (for example, 0.5–1 kg) that is representative of the fraction to be sampled, and then following this method:

1. Measure the weight of oven trays.
2. Measure the combined weight of samples plus trays.
3. Calculate the actual weight of the sample.
4. Place samples in an oven at 105°C for 24 hours and then allow the trays to cool in a dry environment.
5. Weigh the samples and trays.
6. The difference in the weight of the waste before and after drying is the weight of moisture. Now you can calculate the proportion of moisture as a percentage of the total weight.

15.7.6 Market and commercial waste

Markets may create large amounts of waste, including product packaging (paper, plastic), animal slaughter wastes, perished fruit and vegetable wastes, food stall waste, and floor sweepings. Work with market vendors and small businesses to design a waste management plan that deals safely with each type of waste and ensures that public health risks, disease vectors, odours, flies, and other nuisances are minimized. At a minimum each market vendor should have their own 50–100 litre waste container that can be emptied into public waste containers positioned throughout the entire market area. In most settings, paid cleaners will be required to sweep any litter that is created and bring the wastes to a centralized collection point. It is often advantageous to position a large-volume bulk refuse container or construct a transfer station near the market area (or, in fact, in any area that is a large producer of waste).

A high proportion of market wastes are organic in nature and should be collected separately, directly from vendors. In hot and temperate climates, market wastes should be collected daily. It is also preferable for 'wet' mixed waste to be collected separately from 'dry' (paper, plastic packaging) wastes so that they can be disposed of appropriately. Additional infra-structure may need to be constructed in animal slaughter or fish market areas to allow the area to be washed, cleaned, and disinfected. This may require the construction of dedicated soak pits equipped with grease traps (see Chapter 14, section 5.2)

15.7.7 Hazardous wastes

Solid waste providers must ensure that hazardous wastes from domestic or small-scale commercial activities are managed adequately. Examples of these wastes include used motor oils (a source of heavy metals), antifreeze, car batteries (containing lead, mercury, and acid), discarded tyres (containing complex synthetic rubber compounds), used oils (containing various synthetic compounds), asbestos, oil-based paints, paint thinners, wood preservatives, pesticides, and electronic waste (a source of hazardous materials including cadmium, mercury, arsenic, and lead). Great care must be taken as even small quantities of these wastes can cause significant risks. In all cases, steps must be taken to ensure that these wastes do not enter the general mixed domestic waste stream, where they can pose a danger to the health and safety of workers, scavengers, or the environment. Expert advice related to the handling, moving, storage, and transportation of hazardous wastes must be sought when you are faced with toxic or hazardous wastes. Facilities must be provided for the safe collection, bulk storage, and transfer of these wastes to competent authorities where they can be dealt with safely. All hazardous wastes should be clearly labelled during storage and transportation with the name, date, and type of waste in addition to internationally recognized hazard warning symbols. The management of hazardous wastes is complex and expensive, and so a significant element of the budget may have to be allocated to this. Section based on Kayaga & Cotton, 2019 and ** Harvey, et al., 2002.

15.8   Point of generation, storage, and emptying

Solid waste containers are one of the most critical yet most often overlooked elements in any SWM programme. The root cause of many problems associated with solid waste collection can often be traced back to poor waste container selection.

15.8.1 Household containers

Household waste collection containers vary immensely and come in many materials and sizes (Figure 15.1). In most emergency settings it is typically the responsibility of the family to organize its own waste collection at the household level, although the agency responsible for SWM may provide containers, especially if this encourages source separation of wastes. The issuing of household containers in the very early stages of a response is probably unwise; clean new containers (or even plastic bags) have much more important uses than putting garbage in them!

15.8.2 Communal solid waste containers

The choice of communal waste container should be compatible with local practice and with collection and transport mechanisms. It is prudent to assess

Figure 15.1 Selection of household solid waste storage containers

the waste generation rate (see Chapter 15, section 6) and then slightly oversize the waste containers as the cost, time wasted, and public health risk of cleaning up wastes from overflowing containers more than justify the additional cost of larger containers. In addition, containers that are never 100 per cent full reduce the risk of wind dispersion. It is recommended that a safety factor of an additional 100 per cent is allowed to accommodate public holidays and unavoidable service disruptions. An example of calculating a waste bin size is given in Box 15.6.

During the early stages of an emergency response, waste production will be very variable. If no other data exist, those in Table 15.10 are suggested. As soon as possible, undertake a proper assessment of waste accumulation rates and recalculate the optimum waste collection volume for the context. It is essential that the container size and type are compatible with the collection vehicle being deployed. There is no point procuring large waste containers if the location cannot be accessed by waste collection vehicles. Better to provide more, smaller containers that can be accessed by smaller vehicles. Use the largest-sized containers that are compatible with the collection vehicles.

Box 15.6 Optimal waste bin size for an urban collective centre

An urban residential housing block is accommodating 50 refugee families. Waste collection is planned for twice a week.

Assume the daily weight of waste generated is 0.5 kg per person at a density of 300 kg/m³ (Table 15.9).

If the average family size is six, then the total volume of waste generated each day is $0.5 \times 6 \times 50/300 = 500$ litres.

If waste is collected twice a week, the total storage volume required is $500 \times 4 = 2,000$ litres.

Allowing 100 per cent extra for public holidays and service disruption gives a total waste bin volume of **4,000 litres**.

In comparison, Table 15.10 recommends a storage capacity of 2,000 litres but without an allowance of 100 per cent extra.

Source: (** Harvey, 2020)

Table 15.10 Parameters of communal waste storage

Parameter	Value	Comment
Acute phase storage capacity	100 litres per 40 families per day	Daily collection
Stabilization/recovery phase storage capacity	100 litres per 10 families	Daily collection
Storage capacity for established settlements	200–400 litres per 16 families	Low-density, single-storey buildings
	2,000–4,000 litres every 50 m	High-density urban area with multi-storey buildings
Maximum container weight for manual emptying	50–80 kg	Provided vehicle loading height is less than 2 m
Maximum height for storage containers	1 m	Assuming accessibility for children
Maximum walking distance to a communal storage bin	25 m	

15.8.3 Emptying

Manual emptying is common and generates an income for employees but is inefficient, leads to more spillage, and limits the size of container that can be used (Table 15.10). A much safer and more efficient (though also more expensive) primary collection system avoids manual lifting through the use of larger, specifically designed metal containers that can be mechanically lifted onto suitable vehicles (Figure 15.1).

Waste containers should be high enough off the ground to prevent animals (goats, cows, etc.) accessing the waste yet still low enough for children of at least six years old to use. A compromise between these two competing requirements is a target height of 1 metre. Open-sided containers are not recommended: they attract animals to scavenge and scatter the waste around the environment. Containers should have sufficient strength to be able to hold the waste securely without deforming, withstand the loads exerted during loading and emptying, and resist general misuse (knocks and scratches). Wastes in emergency settings are typically up to five times denser than wastes in conventional settings, therefore purchase the most durable containers that are compatible with the collection system. Wet solid waste can become extremely corrosive, so paint steel containers inside and out to reduce corrosion. The bottom of containers should be perforated to allow drainage. In consequence, the floor of the container storage area should be impermeable and self-draining, preferably to a soakaway.

Setting fire to the contents of containers is unfortunately a common practice in many emergency settings, to reduce space in the container, for warmth, for entertainment, or by accident (typically hot ashes are to blame). It is prudent to select waste containers that are constructed from heavy sheet steel and are able to withstand repeated fires. Containers made from wood or plastic, or metallic containers with wheels, should be avoided.

Rainwater should be prevented from mixing with wastes at all costs as it causes wastes to decompose and increases the gross weight, putting additional load on the collection and transportation equipment and staff. The container opening and closing mechanism should be light enough that it can be operated by all users. Unfortunately, covers on communal waste containers are nearly always a failure because people are unwilling to open or replace them (because they are heavy, dirty, or inconvenient). Placing containers in a covered shelter may be an alternative, or use containers with foot-operated lids, such as the one shown in Figure 15.2.

The cost per ton of sweeping up wastes that have been dumped on the floor can be over 10 times as much as the cost per ton of emptying waste that has been placed in a container, and so it is essential that containers are conveniently located and accessible throughout the area. Where the affected population has settled among an established community, it will be necessary to increase the volume of storage provided to accommodate the increased waste generation.

15.9   Solid waste transport

All solid waste programmes involve transferring the wastes from the point of collection to a place of sorting, treatment, or final disposal. In emergency settings, transportation methods can include anything from wheelbarrows to compactor trucks. The cost of moving wastes is often the most expensive part of waste management and the options should be evaluated carefully to select the most appropriate in terms of quality, efficiency, and cost-effectiveness.

Figure 15.2 Foot-operated solid waste storage container suitable for mechanized emptying

15.9.1 Container exchange versus emptying

In some contexts, there may be a choice between vehicles that collect waste emptied into them from storage containers and those that collect full storage containers and replace them with clean, empty ones. The advantage of the exchange system is that there is no on-site transfer of wastes; this reduces the risk of littering and contamination. In addition, bins can be routinely cleaned in a centralized location, reducing the risk of disease vectors and smells. Many medical waste collection schemes use this model. The disadvantage of this system is that waste collection may be less efficient as containers may be taken back half-full and vehicles may be working at less than rated tonnage. In well-controlled settings such as medical centres, containers can be fitted with disposable plastic bags that are replaced at each collection event.

15.9.2 Primary collection vehicle size

The efficient transportation of waste occurs when the capacity of the waste collection vehicle is optimized to the volumes of waste generated and the distances over which it is transported. For each context, it is essential to calculate the total volumes of waste generated and compare the cost per ton per kilometre for each available transport option. In general, the cost per ton per kilometre of waste transported decreases as the storage volume of the transport vehicle increases. Usually, the most cost-effective option is to use the largest primary collection vehicle possible that can physically access the waste collection points. The efficient transportation of waste also occurs when vehicles are loaded to their rated tonnage. Many vehicles (for example, trucks or skip carriers) are designed to carry high-density construction materials and therefore are inefficient for transporting wastes with low densities unless fitted with volume extension panels.

In many high-density settings, it may not be possible for large waste collection vehicles to directly access the waste collection points. Primary collection may have to be undertaken using manual labour or small carts that are able to negotiate the narrow access passageways. In this case, groups of waste containers that can be comfortably and safely lifted or transported by a single person (60–80 litres) should be located within 25 m of every household. Figure 15.3 shows a selection of primary transport vehicle options.

Figure 15.3 Solid waste transport vehicles

15.9.3 Transfer stations

Generally, a transfer station (Figure 15.4) should be installed if a waste collection vehicle is spending more than 15 per cent of its time travelling to and from the disposal site. Transfer stations are usually sited as close as possible to the waste collection area and contain parked bulk refuse carriers and a split-level facility. The split-level setup for transferring wastes is encouraged as it avoids secondary handling and results in efficient gravity transfer without the need for expensive lifting technology. Refuse transfer containers may be over 4 m high and therefore it is worth selecting a site that facilitates the design of a split-level facility and takes advantage of any natural gradients (Figure 15.5).

Transfer stations can also be developed as combined transfer/resource recovery and recycling centres, reducing the amount of waste that ends up entering the landfill. They can also be used as collection centres for common domestic hazardous wastes that must not be allowed to enter the general waste stream.

In emergencies with an affected population of over 20,000, transfer stations should include facilities for staff changing, toilets, showers, an area for servicing vehicles or handcarts, and a manager's office. As the transfer station is typically located within the residential area of the refugee settlement, it is essential that it is kept as clean as possible to reduce odours, vectors, and complaints from residents. The number of transfer stations required depends on the operational range of the primary waste collection vehicles that are being deployed, and the distance to and from the waste-processing area and disposal sites.

15.9.4 Transfer vehicles

Transportation costs make up a significant proportion of waste management programmes. In many cases, it is more cost-efficient for large trucks to make the final journey to the waste disposal site. Bulk refuse carriers typically have a body capacity of up to 40 m³. The most efficient types have removable storage so that multiple container bodies can be filled while the cab and driver are performing haulage.

15.10   Solid waste disposal

15.10.1 Illegal or uncontrolled dumping

This refers to the dumping of solid waste in any location that has not been designated as a disposal site. Although common in emergencies, managers must ensure that it is prevented at all costs as it is a major public health and environmental hazard. The clean-up of uncontrolled dumping is often significantly more costly in terms of time, public health risks, and resources than if the waste had been properly managed in the first place.

1. Loading

2. Transporting

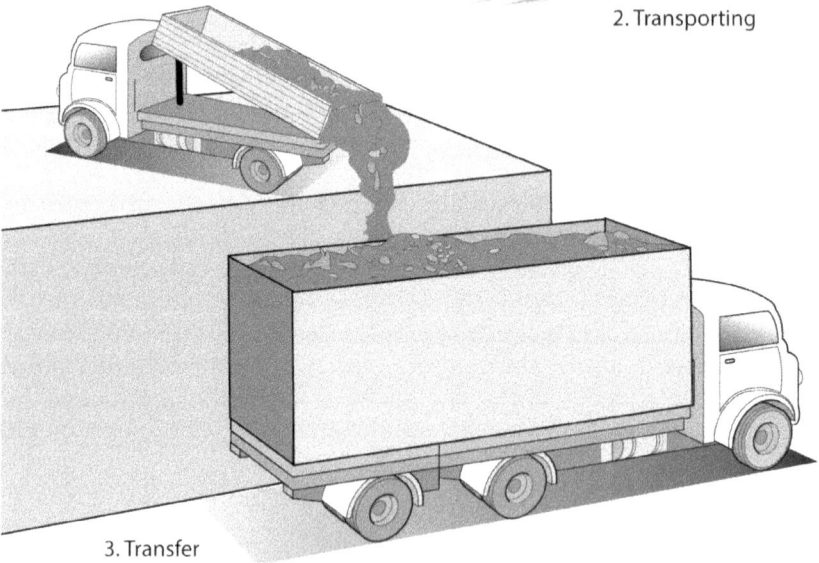

3. Transfer

4. Unloading at disposal site

Figure 15.4 Principles of a transfer station

Figure 15.5 Small waste transfer station

15.10.2 Legal or controlled dumping

Legal or controlled dumping takes place on a piece of land that has been designated by the local authorities as a site where solid waste may be dumped. Even though it may look the same, controlled dumping is better than illegal dumping as it allows a site to be identified with the least possible public health and environmental impact and enables legislative procedures to be implemented to prosecute illegal dumping. It also allows monitoring of the size, types, and volumes of wastes generated and keeps the problem in one location rather than all over the place. Over time, it may be possible to upgrade the site with surface water diversion and daily covering of the waste with soil to prevent vector breeding.

15.11   On-site disposal

In low density communities and as a temporary solution one option is to bury the waste on-site. A hole is dug as far as possible from any building being used by people similar to that shown in Figure 15.9. In the acute phase a pit about 1 metre square and 1–2 metres deep is sufficient but it can be much bigger for longer term use. The pit should be covered and fenced to prevent people falling in, rain entering, and animal scavenging. Cover the waste whenever necessary with a 100–150mm layer of soil to reduce fly breeding and bad odours. When the pit is filled to 500 mm of the ground level it should be filled in with soil to slightly above ground level and left undisturbed, preferably for at least two years.

15.11.1 Sanitary (engineered) landfill

Sanitary landfill is the process of disposing of solid waste in such a way that it causes minimum impact to the environment (Figure 15.6). Once a landfill site has been filled, it is possible to put it to use for recreation, agriculture, construction, or other purposes. The landfill location is usually carefully selected to have the best hydrogeological conditions to avoid surface and groundwater pollution. Typically, the site has either soils with a high clay content or an impermeable rock base, but in permeable soils the base of the site may have to be sealed with a strong, long-lasting, impermeable membrane or clay lining. The site should also be a long distance from groundwater and surface water sources. In a sanitary landfill, animal and insect vectors are eliminated by a daily covering of waste with 10–20 cm of clay soil. This also serves to kick-start biological decomposition and to divert rainwater, reducing the amount of leachate that is produced within the waste. Over a period of years, the waste decomposes through natural biological and chemical processes (principally anaerobic decomposition) to render the wastes inert. Sanitary landfill is a simple approach that requires no imported technology and is well suited to emergency settings.

A common misconception is that landfill sites can be used to dispose of any wastes. This is false: they should receive only wastes that have been deemed 'acceptable' for the site. Hazardous wastes must not be allowed to enter the domestic waste stream and must not enter the landfill as they will not biodegrade and may contain long-lasting and highly polluting elements.

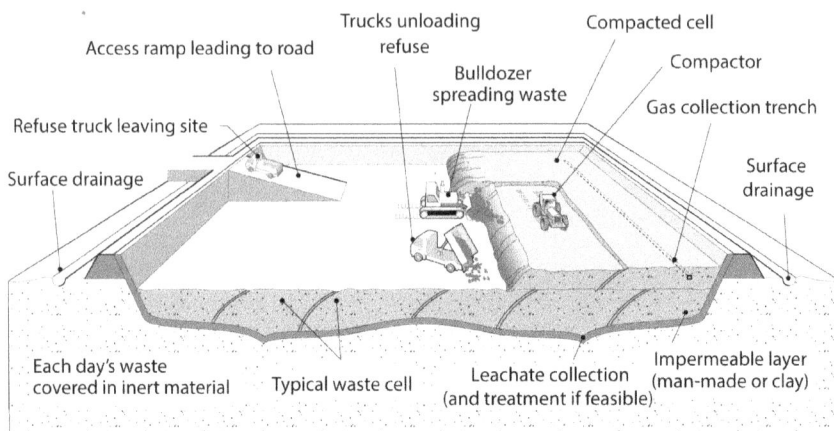

Figure 15.6 Sanitary landfill site

Sanitary landfills produce liquid effluent called leachate that is highly polluting because of its chemical and biological content. If the leachate is contained within the landfill, it will slowly degrade through a number of biological processes. Alternatively, the leachate can be channelled away from the site and treated separately as a high-strength wastewater using technologies described in Chapter 9. Following good sanitary practice will minimize contamination from rain and stormwater, which, in turn, will reduce the quantity of leachate. The natural anaerobic decomposition also produces copious quantities of gases, including methane, carbon dioxide, hydrogen sulphide, and ammonia. These can be explosive and should be prevented from building up by installing vent pipes (not shown in Figure 15.6). Smoking on-site must be prohibited.

Land requirements
Making even an approximate estimation of the space required for landfill is necessary when selecting a new site or adding to an existing one. In the early stages of an emergency response, accurate data will be unavailable, but even a rough approximation of requirements will be helpful. Box 15.7 provides an example of how to estimate space requirements.

Box 15.7 Calculating the land required for a long-term sanitary landfill facility

A permanent landfill site is required for a refugee camp of 60,000 people. A detailed waste audit has shown that the landfill fraction of waste is currently 0.3 kg per person per day. A permanent landfill location is required with a design life of at least 10 years. What area of land needs to be found and developed?

Projected Pop (P) = Pop/$(1 - R)^N$
Where:
Pop = the current population (60,000)
R = the population growth rate (3%)
N = the number of years (10 years)
Therefore:
Projected Pop at the end of 10 years (P) = $60,000/(1-0.03)^{10}$ = 81,364 persons
Average lifetime population = (60,000 + 81,346)/2 = 70,682 – say, **71,000**

The volume of landfill waste that will be accumulated is as follows:
Volume (V) = [N × P × R × 365 × F]/ρ
Where:
N = effective life of the landfill (10 years)
P = the average population (71,000)
R = waste accumulation rate (0.3 kg/person/day)
F = factor for daily soil cover (2)
ρ = compacted waste density (500 kg/m³) (see Table 15.9)
Therefore:
Volume = [10 × 71,000 × 0.3 × 365 × 2]/500 = 310,979 m³– say, **310,000 m³**

Assuming a flat site and a landfill total height of 4 m gives a landfill area of:
Landfill area (A) = volume/depth
Landfill area (A) = 310,000/4 = **77,500 m²**

(Continued)

> **Box 15.7** Continued
>
> Assuming a square site, the dimensions of the landfill site will be:
> Length of each side (L) = Area (A)$^{0.5}$ = 77,500$^{0.5}$ = 278.39 m – say, **280 m**
>
> Depending on the shape of the land, additional space will be required for offices, parking, vehicle maintenance, turning, etc. On a site this size, an additional 25 per cent is suggested.

> **Box 15.8** Sanitary landfill site selection criteria
>
> - **Hydrogeology:** The geology of the ground should minimize the risk of leachates passing directly into the groundwater or surface water.
> - **Cover material:** The site should have a large quantity of clay or clay loam soil available for daily cover material. Where clay soil is not available there will be a larger volume of leachate generated.
> - **Distance:** For efficiency, the site should be as close as possible to the affected population but not so close that it creates a serious hazard from odour, vermin, flies, etc. There is no fixed minimum distance recommended – it is one of the factors to consider when selecting a site.
> - **Access:** The condition of access roads must be capable of handling the quantity and weight of waste haulage traffic.
> - **Topography:** There should be no surface water features running through the site and no surface water bodies downstream.
> - **Nuisances:** The site should be downwind from residential areas.
> - **Potential volume:** The site should have a sufficient space for at least 10 years of use.
> - **Permission:** Just because it is an emergency doesn't mean the land isn't owned by someone. Generally, land owned by the government is easier to access than private land.
> - **Regulation:** Many countries have very strong regulations governing the siting of landfill sites. Governments may be willing to set some aside as a temporary measure, but negotiations can take a long time.
>
> *Source*: Adapted from (** Harvey, 2020)

Identifying a suitable location for a landfill

Site selection for a sanitary landfill can be a controversial activity and must be carried out in full cooperation with the landowners, local authorities, and local population. The process of site selection is generally carried out by considering a number of potential sites (Box 15.8) and selecting the best available site based on minimizing risk. Obtaining formal approval for a new landfill site nearly always takes many months, if not years. It may be possible to establish a temporary site with strong commitments to rehabilitate the site at the end of the emergency, but it is nearly always better to make use of an existing site rather than try to establish a new one.

Preparing a landfill site

How the landfill site is prepared depends primarily on the topography and dimensions of the site. Typically, the landfill is organized as a series of cells arranged using either the trench method or the area method (Figure 15.6). A master plan for the site is prepared that divides the site into cells, starting at the lowest elevation and the point furthest from the entrance. On sloping land, the cells should follow the contour of the land. The width of the cells depends on the number of vehicles that are likely to offload their waste at the same time during peak operation. In terms of management, it is best to have the smallest operating face possible; even in large emergencies, it is unlikely that more than one waste collection vehicle will be arriving to offload wastes at any one time. A cell width of 10 m should be sufficient.

An important factor when managing the landfill is that there are sufficient quantities of daily soil cover material. At least 1 m^3 of soil cover material is required for every 1 m^3 of waste that is deposited. The primary source for this cover soil is the waste cell itself. The cover soil must contain a high proportion of clay in order to prevent rainwater and surface water from entering the landfill and creating additional leachate. Finally, methane gas venting pipes must be installed to stop the build-up of explosive gases under the cover soil.

On very small sites, site preparation, waste spreading, compaction and backfilling may be undertaken manually provided adequate PPE and suitable tools are provided. In most cases however the work will be done by machines. The most cost-effective solution for a large site serving up to 80,000 persons is to use a four-wheel drive tractor equipped with a wide front loader bucket, a rear digging bucket (backhoe), and a grader attachment. Several 6-ton capacity two-wheeled tipping trailers are also useful for moving large quantities of soil cover material. The site should have an enclosed garage, fuel storage and the cost of servicing and backup should be budgeted.

Decommissioning a landfill
When each cell within the landfill has reached capacity and is due for decommissioning it should be completely capped with 0.5m of soil and landscaped back to its natural appearance. Plants, including grasses, may be planted into the final capping layer to stabilize the soil and reduce leachate quantities through evapotranspiration. It takes many years for a landfill site to be stable enough to support buildings without expensive foundations. The site will experience subsidence as well as gas emission and therefore will probably need to be left as open land for some years. It may be possible to return the land to agriculture if appropriate topsoil is added but sometimes the gases given off will poison the crops. In urban areas careful monitoring of leachate and gas discharge is required but it may be possible to use the site as a temporary car park.

15.11.2 *Waste incineration*

Incineration is high-temperature combustion under controlled conditions. However, it is very unlikely that an incinerator would be constructed to dispose exclusively of solid waste from emergency settings (except for the disposal of medical waste (See Chapter 15, section 13.1) because of the high cost and long construction period. It may be appropriate to transport waste to an existing incinerator but the calorific value of waste for emergency settings tends to be lower than normal waste and it may not be accepted.

The open burning of waste at a disposal site is extremely dangerous to public health and can produce harmful or carcinogenic substances, including dioxins. It should never be approved.

15.11.3 Composting

Composting is the biological decomposition and stabilization of organic material, such as vegetable scraps, under aerobic conditions (in the presence of oxygen). Under the correct conditions of moisture and aeration, heat is generated and composting takes place. The rise in temperature is also sufficient to kill pathogenic organisms, making the final product safe to handle. The composting period is followed by a period of stabilization to produce a final product suitable for application on the land without adverse environmental effects.

Any organic material can potentially be composted. However, woody materials such as woodchips and paper take much longer to compost than fleshy materials such as vegetables and vegetable peelings. It will, however, be useful to get a good balance between woody materials and fleshy organic material. Woody materials are important for keeping the structure of the compost open and to allow air in, thereby keeping the composting process aerobic; this speeds the process and keeps odours to a minimum. The fleshy organic material will be the main matter to be composted. As woodchips don't compost quickly, these can be sieved and reused for future composting if required.

Bin composters
Compost bins are most suitable for use in single households or small communal environments to compost vegetable waste and garden cuttings. The bins can be bought purpose built or can be easily constructed using an ordinary household bin with holes penetrating the sides to enable air to circulate within the contained compost. A box compost container can also be constructed using wood planks to form a slatted box container (Figure 15.7). Vegetable peelings and kitchen waste are added to the compost bin and left to 'compost'. Woody and fibrous materials can be added; these assist with maintaining airflow through the compost, but they will take longer to break down.

One of the main problems with compost bins is that they do not support true composting. The amount of material and the rate at which it is added to the bin are insufficient to raise the temperature enough for true composting to take place. It can take a considerable period of time for the organic material to fully break down (up to a year). For this reason, faecal matter should never be added to bin composters. Also, restrictions on the amount of air in the system can lead to anaerobic conditions, which can create bad odours and attract flies and vermin. The most appropriate way of dealing with this is to turn the

Figure 15.7 composting bins

compost occasionally. Cover the bin with a lid to reduce access to the compost by vermin as well as keeping out excessive rainwater.

Wherever possible, a twin composting system is recommended. This involves utilizing two compost bins: one compost bin that is in use and being filled with new vegetable waste; and another compost bin containing older 'composting' waste that is no longer being added to. This system has the advantage that the bin containing the composting material can be turned without new, un-composted matter being added and slowing down the overall composting process.

Windrow composting
Windrow composting offers a relatively low-technology, low-cost composting option for large quantities of organic waste. Typically, organic wastes with a high water content (e.g. vegetable waste) are blended with a drier material such as straw, woodchips, or mature compost. Doing this opens up the compost mass, encouraging air movement and enabling the compost to be maintained in trapezoidal or triangular section windrows around 1.5 m in height (no greater than the average height of a compost worker), as shown in Figure 10.11.

The large volume of organic material supports composting provided that the moisture content is maintained and the pile kept aerobic. Aerobic conditions are maintained by regularly turning the mass so that the outside of the pile is turned into the middle and vice versa. The frequency of turning is dependent on the temperature of the mass. The composting process only takes place when the mass is between 60°C and 32˚C. The windrow should be turned when the temperature of the centre of the mass goes above or below these figures. This may be as frequently as every one or two days in the beginning but will reduce to around weekly after about 30 days. Water can also be added during the turning process if the pile becomes too dry. Composting is complete when the pile no longer heats up even after turning. The compost should then be stockpiled for four to eight weeks to ensure that it is stable and does not further degrade when added to soil. Faecal material should not be added to windrows until the operators are proficient in operating the composting process and are sure that the final compost is sterile.

Ideally, composting should take place on a concrete base with cut-off drains around the edges to capture leachate or runoff (from rain); this avoids environmental degradation of the surrounding area, watercourses, and groundwater. Leachate can be captured and re-added to compost during the turning process to add moisture to the pile if drying is an issue. Piles may need to be covered in areas with high rainfall or in hot climates to reduce evaporation.

Windrow composting should be undertaken in batches with a single windrow containing three to five days' volume of organic waste. Care must be taken to ensure that batches are not mixed; mixing will slow the overall composting process and may lead to incomplete sterilization.

Windrow composting can be highly mechanized, but if it is to be undertaken using hand tools, the piles should be kept small for ease of turning (Section adapted from Oxfam, 2008).

15.12 Operations management

The design and technology around SWM are relatively simple; the problem lies in its management. SWM employs large numbers of very low-skilled workers who tend to be poorly paid and have low esteem within the community. In addition, the opportunity for low-level corruption and petty theft is great. Management consists of a daily mix of personnel problems, mechanical breakdowns, and complaints from customers. It requires close monitoring of all aspects of the waste stream and good relations with staff. Good public relations and communication skills are also needed because efficient and effective SWM cannot be achieved without community support.

15.12.1 Workforce management

In any organization, the employees are the most valuable resource. The role of the manager is to ensure that the people for whom they are responsible can perform their duties in the most effective and economical way. To achieve this, the manager must consider the following aspects:

- **The tasks assigned:** Do the employees know what they are expected to achieve and how to do the work?
- **Tools, equipment, and PPE:** Do the employees have the tools, equipment, and PPE needed to do the work effectively, efficiently, and safely? Do they know how to use the equipment and how to get it repaired or replaced when necessary?

- **Motivation and supervision:** In general, the payment of a salary alone does not motivate a worker to do their best work (though a very low salary can have a negative effect on motivation). The best motivation results from pride in work well done and a sense of responsibility. However, it is necessary to encourage good standards of work by means of fair supervision and the imposition of unwelcome consequences for inadequate performance. Foremen and supervisors therefore play an important role.
- **Communication should be two-way:** Not only should instructions be handed down from superiors, but suggestions and complaints from junior staff should also be accepted and considered by superiors. Many good and practical ideas can originate from those carrying out the task.

Training

Staff training in SWM is critical to the efficient and effective operation of the process but has often been unsatisfactory for the following reasons:

- The training material is unsuitable for the particular situation. Issues such as legislation, waste characteristics, and local practices vary almost infinitely.
- The wrong people are trained. In many countries attendance at training courses is seen as a perk and therefore assigned to senior staff who do not benefit from the course.
- Training methods are not suitable. SWM is primarily a practical activity carried out by relatively poorly educated individuals for whom theoretical training is unsuited. Practical-based training is much more effective.

Supervision

Supervision is very important in ensuring that the work is done in a correct and efficient way. Foremen and supervisors should understand that they themselves are being supervised and are accountable for the way they do their work. In a large city, there may be many levels of supervision. Often there is a foreman (or chargehand) who supervises four to 20 labourers. Groups of four to 10 foremen/women are supervised by supervisors, who are in turn supervised by district superintendents. Job titles and numbers vary according to local conditions. Supervisors need a means of transport and some form of communication equipment in order to be able to work effectively.

Supervising drivers

The supervision of drivers of waste collection vehicles can be a particular problem. If there are separate sanitation and transportation departments, and the driver is under the transportation supervisor, the sanitation supervisor is not able to instruct them directly regarding collection procedures. This may affect the whole collection crew. If the driver is under the sanitation supervisor, then the transportation supervisor cannot instruct them directly regarding the use, care, and maintenance of the vehicle. In some situations, it may be desirable for the driver to also act as the foreman of the collection labourers. This cannot be done if the driver is in a different department.

15.12.2 Working with government

In most countries, waste management is devolved to a local authority such as a municipality or district council. Emergency support staff should not undermine or replace the local authorities or their dedicated contractors responsible for SWM; rather, they should carry out SWM in close collaboration with the national providers and in full compliance with solid waste and environmental legislation. Where possible, existing services should be extended to the emergency-affected population. In urban settings, local authorities or service providers responsible for SWM may be overwhelmed by the additional burden created by the emergency and may require support (additional equipment, staff, training, and resources) to bring back service levels to the same as before the emergency.

15.12.3 Private sector participation

Private sector participation is widespread in the SWM industry, from using self-employed waste collectors to managing the whole process. As with any management option, it has its advantages and disadvantages:

Advantages
- **Specialist expertise:** Large private companies that specialize in SWM are likely to have a considerable pool of experts with relevant specializations who are well qualified to set up sustainable systems.
- **Access to capital:** The private sector often has much better access to capital (from commercial loans and reserves), which enables it to buy the most efficient equipment and establish facilities such as landfill sites that meet the required specifications.
- **Freedom from bureaucracy:** The profit orientation of private companies obliges them to operate as efficiently as possible, so they are not hindered by bureaucratic processes in the same way as many local government administrations. As a result, for example, vehicles can be repaired in a much shorter time.
- **Clear financial arrangements:** Agreements and contracts specify precisely the level of financial commitment that is involved, so municipal planning and budgeting can be based on known amounts.
- **Control:** If contracts are well written and carefully enforced, there can be more control over the operational and environmental performance than if the service is provided by local government. This is especially true if environmental monitoring by the relevant government agency is weak.

Disadvantages

- **Complex contracts:** Preparing a contact requires a high level of expertise and an advanced knowledge of what is required from the contractor in terms of the level of service, the charges to be made, and the outputs expected over the whole of the contract period.
- **Apparent cost:** At the stage of transition from public sector to private sector provision, it may appear that the private sector is more expensive because the total expenditures for public sector service provision are not known.
- **Corruption and fear of accusations:** Involving the private sector does create new opportunities for corruption. In some cases, the fear of being accused of corruption causes greater problems, as officials exploit a contractor harshly in order to persuade others that they are not corruptly favouring them.
- **Loss of capacity:** Organizations that engage the private sector to provide services may lose the capacity to provide the service themselves if the private sector operator fails or demands too high a fee. In larger communities, the public sector may continue to provide services in some of the area, in order to maintain the ability to take over from the private sector operator if necessary.
- **Transfer of workforce:** In many cases, labourers and drivers who work for local government have better pay and conditions than are found in the private sector. If the work is given to the private sector, the public employees may either find themselves without a job or be obliged to work for less pay and fewer benefits. Private sector organizations that are taking over the work may be reluctant to hire public sector workers who have developed bad working habits because of overstaffing and poor supervision.
- **Poor service delivery:** A private sector provider is only as good as those who supervise it. Poor management expertise within the lead organization will lead to poor-quality serve delivery.

15.12.4 Engaging with the community

In identifying and developing SWM options, it is important to involve communities from the earliest possible stage. They can assist with identifying those who have previously been involved in the waste sector and have knowledge of existing recycling markets. As with all activities where community groups are consulted, it is important to obtain the views of different sections of the community, as they all have different needs. Men, women, children, and those with particular or special needs should be consulted and considered separately.

In undertaking recycling and composting activities, separation of waste at source will be important to success. Source separation involves waste being separated by those who generated it. It has an advantage over separating later, as it avoids double handling and unnecessary transportation.

Options for separation at source may include a central location where people can bring their recyclable materials to sell, which are then sold on to commercial recyclers and community recycling points where people separate the waste that is then taken away by a central coordinator and sold to commercial recyclers.

Determining the most appropriate option depends on what people are familiar with doing, and whether formal or informal recycling markets already exist and how accessible they are to the local population.

To enable recycling and composting to take place successfully, communities need to understand why it is important to separate waste and how they can do this in a practical way. The role of the public health promotion team is crucial in understanding current practices and in raising awareness of new issues related to better waste management practices. The roles of men, women, and children in any potential recycling and composting activities should also be analysed, as these are likely to have a big impact on potential outputs.

Identifying individuals to champion recycling and composting within communities can assist with presenting messages and also with linking the management of waste to broader health messages (Oxfam, 2008).

15.13   Medical waste

Medical waste is generally defined as all infectious and non-infectious waste material arising from healthcare-related activities. Medical waste collection services must comply with World Health Organization protocols (WHO, n.d.) and ensure that healthcare wastes are segregated at the point of generation according to four categories:

1. sharps (needles, scalpels, etc.), which may be infectious or not;
2. non-sharps infectious waste (anatomical waste, pathological waste, dressings, used syringes, used single-use gloves);
3. non-sharps non-infectious waste (paper, packaging, etc.);
4. hazardous waste (expired drugs, laboratory reagents, etc.).

Steps should be taken to ensure that waste is segregated and treated appropriately without risk to staff, patients, the public, or the environment regardless of the emergency setting or type of healthcare facility.

All healthcare waste should be collected in dedicated colour-coded and labelled containers according to its contents for safe handling (Table 15.11). Colour-coding facilitates easy identification and segregation on the basis of waste hazard classification and suitability of treatment and disposal. Colour-coding also makes the segregation process understandable for low-skilled workers with limited literacy. It is essential that the adopted colour-coding system is used consistently throughout the healthcare waste management chain to avoid confusion and mismanagement of the waste.

Table 15.11 European waste catalogue colour-coding for healthcare waste containers

Waste type	Classification	Colour-coding	Description and disposal method
Infectious	Hazardous	Yellow	Infectious waste that requires disposal by incineration.
Infectious	Hazardous	Orange	Infectious waste that may be treated to render it safe prior to disposal. Alternatively, it can be incinerated.
Cytotoxic/ cytostatic	Hazardous	Purple	Waste consisting of, or contaminated with, cytotoxic and/or cytostatic products that requires disposal by incineration.
Offensive	Non-hazardous	Yellow and black	Non-infectious, offensive/hygiene waste that may be recycled, incinerated, or deep landfilled.
Anatomical	Hazardous	Red	Anatomical waste that requires disposal by incineration.
Medicinal	Non-hazardous	Blue	Waste medicines, out-of-date medicines, and denatured drugs that require disposal by incineration.
Dental	Hazardous	White	Dental amalgam and mercury to be stored safely to prevent exposure to mercury vapour and require disposal by recovery or recycling. This includes spent and out-of-date capsules, excess mixed amalgam, and contents of amalgam.
Domestic	Non-hazardous	Black	This waste should not contain any infectious materials, sharps, or medicinal products and requires disposal by landfill.

Source: Adapted from (Initial, 2021)

15.13.1 Medical waste disposal

Due to the risk of scavenging, ensure that medical waste is managed separately from the general waste streams. A dedicated medical waste disposal zone should be established as close to the healthcare facility as possible and at least 50 m (preferably downhill and downwind) from any buildings or public areas. The site should be fenced off and locked. All pits should also be locked and covered. The waste disposal zone should have a water point with soap or detergent and disinfectant for handwashing and for cleaning and disinfecting containers and tools. There should also be an apron to catch the wastewater, with disposal into a soakaway or sewer. Ideally, there should also be a storage area where tools and containers can be dried and stored. The waste disposal zone should also be at least 30 m from groundwater sources, with the base of any pits at least 1.5 m above the groundwater table. Where an incinerator is used, it should be located to allow for effective operation and with minimal air pollution of the health facility, nearby housing, or crops. The site should also be large enough for extension if new pits or other facilities have to be built. Surface water runoff should be managed to prevent entry to the waste disposal area.

Sharps disposal

Sharps should be placed immediately in puncture-proof and covered safe sharps containers (usually yellow or orange) that are collected regularly for disposal. Sharps are potentially the healthcare waste item that represents the most acute hazard in a healthcare facility. Hypodermic syringes should not be manipulated or separated from their needles after use and sharps boxes should be large enough to accept hypodermic syringes complete with their needles. If puncture-resistant containers are not available, the healthcare facility may choose to adapt existing puncture-resistant containers, such as empty water, oil, or bleach bottles made from sturdy plastic. Waste containers for sharps should be available within arm's reach of where they are used. Sharps should be disposed of in a sharps pit (buried drums in small health centres or emergency structures; concrete-lined pits in other settings) (Figure 15.8). Incineration is not recommended in the acute and stabilization phases of an emergency because of the difficulty in achieving the necessary temperatures to fully destroy the sharps. Off-site treatment in an existing centralized facility may be possible in an urban area where facilities exist and on-site treatment or disposal is not feasible because of lack of space.

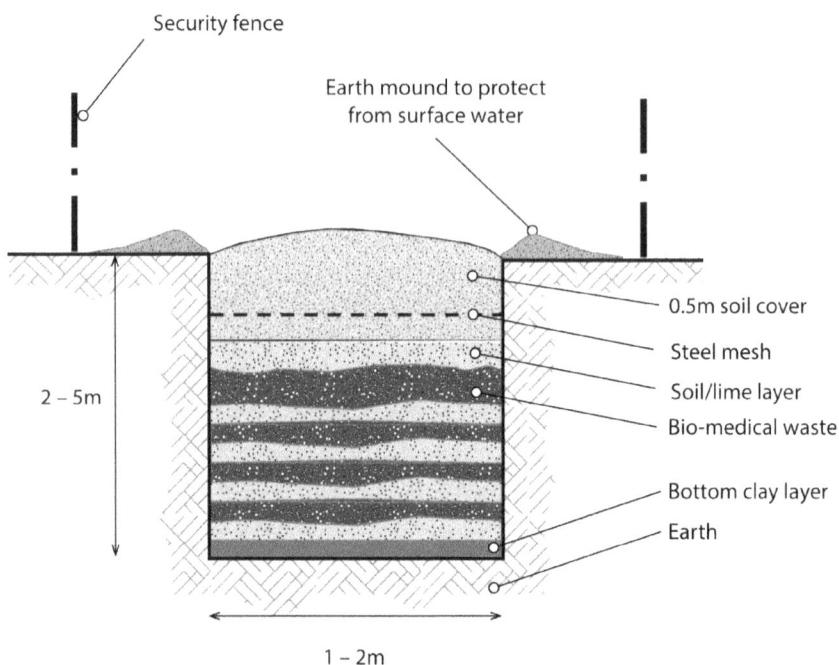

Figure 15.8 Temporary sharps pit

Non-sharp infectious medical wastes

Non-sharps infectious wastes should be placed immediately in the appropriately coloured bags or containers (15–40 litre capacity, with lids). Containers should be collected, emptied, cleaned, disinfected, and replaced after each intervention (e.g. each delivery in a maternity unit) or twice daily. They should be available within easy arm's reach of all locations where these wastes are generated. Waste containers should be transported carefully to avoid spillage. Bags should be delicately handled by the neck, to prevent tearing and spillage. Double bagging may be necessary in some cases. Any leakages or spillage should be cleared up according to infectious waste handling protocols for the clean-up of blood and other body fluids. In the acute and stabilization phases of an emergency, non-sharps infectious waste should be buried on-site in a pit fitted with a sealed cover and ventilation pipe (Figure 15.9). Arrangements may be needed for disposing of placentas, according to local custom.

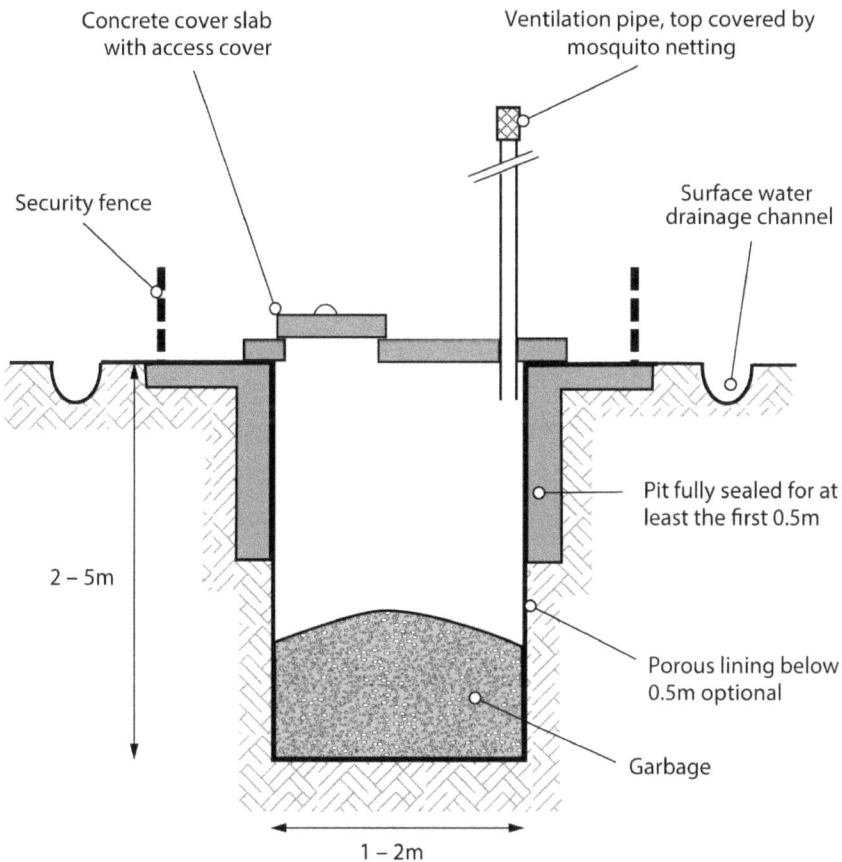

Concrete cover slab with access cover

Ventilation pipe, top covered by mosquito netting

Security fence

Surface water drainage channel

Pit fully sealed for at least the first 0.5m

2 – 5m

Porous lining below 0.5m optional

Garbage

1 – 2m

Figure 15.9 Non-sharps disposal pit

Non-sharp non-infectious medical wastes
Non-sharps non-infectious black waste containers (20–60 litre capacity) should have fitted lids and should ideally be operated with a foot pedal to prevent contaminating hands. Waste containers could be lined with black-coloured sacks to simplify emptying and transport. Containers should be emptied, cleaned, and replaced daily. Temporary covered waste collection containers (e.g. buckets with covers, plastic drums, etc.) may be used for the short term in an emergency provided that they are durable, adequately labelled, and sufficiently leak-proof to safely contain the waste from the point of creation to the point of disposal. It is recommended that there is at least one waste container in every room and at least one set of waste containers per 20 beds in a ward.

Non-sharps non-infectious waste should be buried in a pit or a landfill site. If space is limited, non-sharps non-infectious waste could be incinerated in a low-temperature incinerator; in emergency settings, this may be as simple as an oil-drum type incinerator (Figure 10.13). Ashes and residues should be buried in a pit. Some wastes, such as pharmaceutical wastes, cannot be disposed of in this way and should be sent to a large centre for destruction or returned to the supplier. In all cases, national legislation should be followed. Where longer-term use is expected, a dedicated incinerator can be constructed similar to the one shown in Figure 10.14.

This chapter is largely based on two documents (** Harvey, 2020) for camp-based settings and (** UNEP/OCHA Envionment Unit, 2011) for urban disasters.

APPENDIX 1
Soil characteristics

Table 4.3 gives a rough approximation of the effluent infiltration capacity of different soils. If a more accurate assessment of infiltration capacity is required, there are a number of alternatives.

Soil texture assessment

Table 4.3 also uses soil texture to categorize different soil types. There is further information below on how to carry out such an assessment.

For an infiltration pit, dig or drill a hole to the full depth of the pit; for an infiltration trench system, dig a number of holes at various places around the area of the proposed system. Take a number of soil samples at different depths and locations around the infiltration site. Remove any organic or man-made material and large stones from each sample.

Place around three teaspoonfuls of each sample in the palm of the hand and add a small quantity of clean water to make the sample moist but not so wet that some of the sample runs off the hand. Squeeze the sample into a ball and then follow the chart shown in Figure A1.1. Using this method, an overall picture can be produced of the soil type and hence the infiltration capacity.

The sedimentation method

This method uses the fact that larger soil particles generally settle quicker in water than smaller ones.
Method

1. Collect soil samples as described above. Remove organic and man-made materials. Large stones should be removed if there are only a few; otherwise they must be crushed.
2. Place a sufficient quantity of the sample in a clear container to fill about one-third of the container volume.
3. Add a similar quantity of clean water together with a small quantity of detergent.
4. Cover the container with a close-fitting lid and shake the contents thoroughly.
5. Place the container on a flat surface and leave to stand for 24 hours. By that time most of the soil particles will have settled out, with the larger particles at the bottom and the finer ones at the top (Figure A1.2).
6. Measure the total thickness of settled material (in millimetres) and the thickness of each different layer.

Does the soil lack all cohesion? — Yes → **Sand**

No ↓

Is it difficult to roll the soil into a ball? — Yes → **Loamy Sand**

No ↓

Does the soil feel smooth and silky as well as gritty? — No → **Sandy Loam**

Yes → **Sandy Silt Loam**

START HERE Rub moist soil between thumb and fingers

Does the moist soil feel or sound predominantly rough and gritty? — Yes

No → Does the soil mould to form an easily formed ball and feel smooth and silky? — Yes → **Silt Loam**

No ↓

Does the soil mould to form a strong ball which smears but does not take a polish? — Yes → Is the soil also rough and gritty? — Yes → **Sandy Clay Loam**

Is the soil also smooth and silky? — No → **Clay Loam**

Yes → **Silty Clay Loam**

No ↓

Does the soil mould like plasticine, polish and feel very sticky when wetter? — Yes → Is the soil also rough and gritty? — Yes → **Sandy Clay**

Is the soil also smooth and buttery? — No → **Clay**

Yes → **Silty Clay**

No ↓

START AGAIN or soil is organic

Textural Group	Textural Class
Sands	Coarse Sand / Sand / Fine Sand / Loamy Coarse Sand
Very Light Soils	Loamy Sand / Loamy Fine Sand / Coarse Sandy Loam
Light Soils	Sandy Loam / Fine Sandy Loam / Sandy Silt Loam / Silt Loam
Medium Soils	Sandy Clay Loam / Clay Loam / Silty Clay Loam
Heavy Soils	Sandy Clay / Clay / Silty Clay

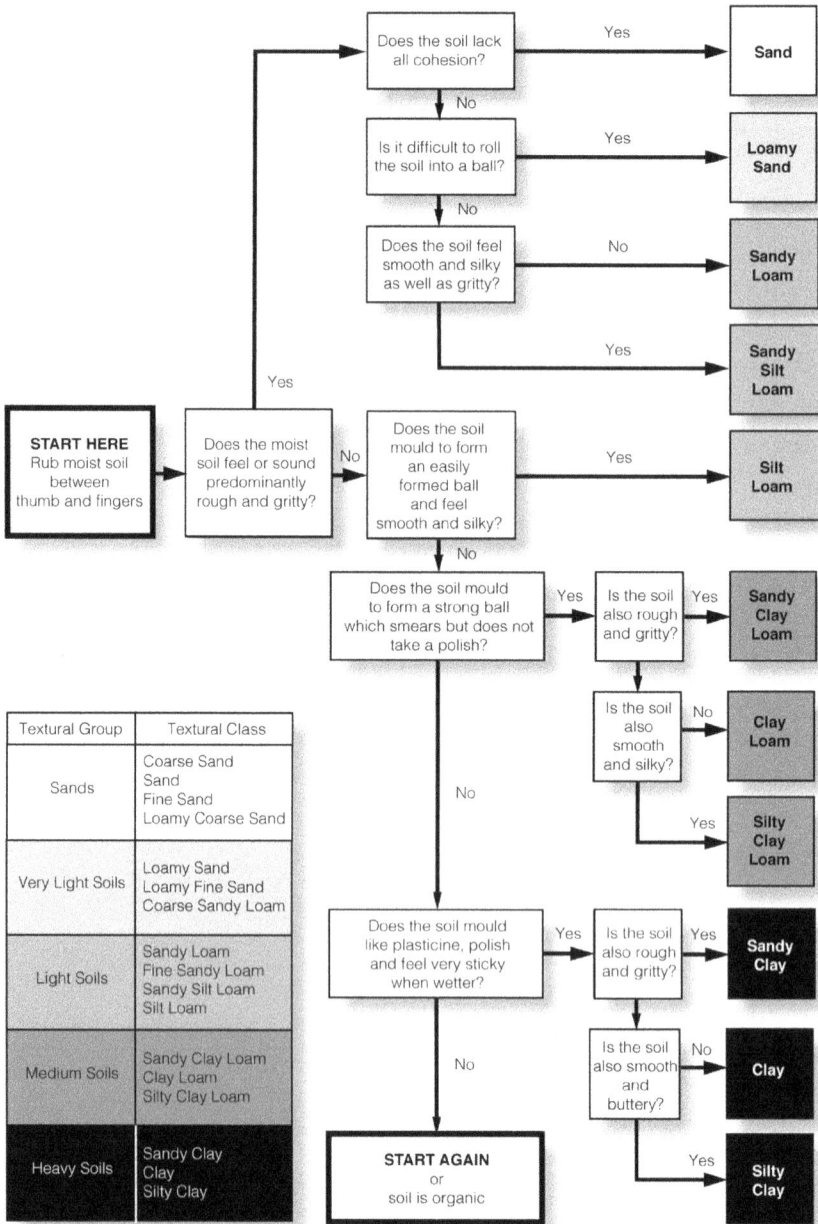

Figure A1.1 Guide to assessing soil texture
Source: Adapted from (Ciba-Geigy Agrochemicals, 1986)

7. The percentage of each particle type can be found by dividing the thickness of the individual layer by the total sediment thickness and multiplying by 100.
8. Compare the results with those shown in Figure A1.2 to assess the soil type.

Sandy soil	Clay soil	Loam soil

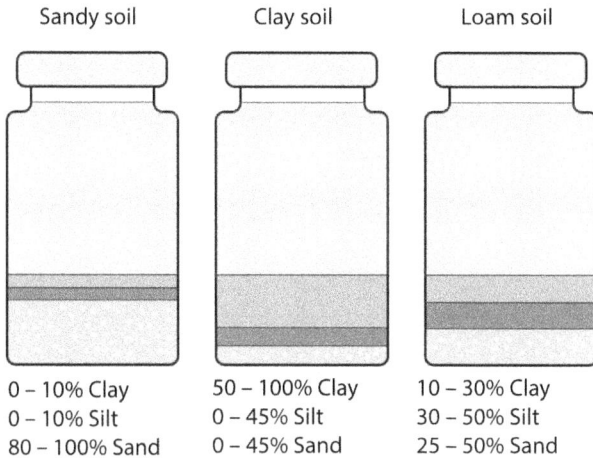

0 – 10% Clay	50 – 100% Clay	10 – 30% Clay
0 – 10% Silt	0 – 45% Silt	30 – 50% Silt
80 – 100% Sand	0 – 45% Sand	25 – 50% Sand

Figure A1.2 Soil sedimentation test
Source: Adapted from (Deep Green Permaculture, 2009)

Percolation test

The percolation test measures the rate at which clean water can percolate into the soil. There are many such tests but the one described is probably the simplest:

- Dig a hole in the ground in the area where the infiltration pit or trench is to be constructed. The hole should be 300 mm square and extend about 300 mm below the top soil.
- Place a sheet of plastic over the base of the hole to prevent infiltration through the base and weigh it down with the minimum quantity of gravel to prevent it moving during the test. Once complete, measure the dimensions of the hole accurately, measuring the depth from the top of the plastic sheet.
- Fill the hole with clean water up to the base of the top soil and keep it topped up for four hours. This is to saturate the surrounding soil.
- Place a ruler vertically in the pit and fill the pit with water once again to the base of the topsoil and note the depth of water.
- Measure the water depth at regular intervals (15, 30, and 60 minutes) until the rate at which the water level drops becomes constant.
- For each time interval, calculate the infiltration rate as shown in Box A1.1.
- Repeat the test to see if the results are the same.
- Repeat the test in other places in the infiltration area if the soil type is variable.

Box A1.1 Calculating the infiltration rate

Assume pit dimensions 300 mm square & 300 mm deep below topsoil therefore pit perimeter length (metres) = 1.2

Test results	Row							
Time (mins)	1	0	15	30	60	120	240	360
Water depth (mm)	2	300	297	296	294	292	289	285
Average water depth (mm)	3		298.5	296.5	295.0	293.0	290.5	287.0
Infiltration area (m^2)	4		0.36	0.36	0.35	0.35	0.35	0.34
Volume of water infiltrated (litres)	5		3.6	1.2	2.4	2.4	3.6	4.8
Infiltration/m^2 (litres/m^2)	6		10	3	7	7	10	14
Infiltration rate (litres/m^2/day)	7		965	324	325	164	124	167

	Description	
	3	Average water depth = sum of starting depth + finishing depth divided by 2
	4	Infiltration area = average water depth × pit perimeter length
	5	Volume of water infiltrated = drop in water level since last measurement/1000 × pit perimeter length
	6	Infiltration per square metre = Volume of water infiltrated/infiltration area
	7	Infiltration rate per day = infiltration rate per square metre × 60 × 24/time over which the water drop occurred

Conclusion: Ignoring the high rates for the earlier results (probably due to influence of topsoil) the infiltration rate can be assumed to be around **150 litres/m^2/day.** Extending the test longer would probably give you a more accurate result. This makes the soil a clay loam having a wastewater infiltration rate of **10 litres/m^2/day**

APPENDIX 2
Chlorine solutions

Chlorine is a readily available and widely used chemical that is highly suitable for disinfecting surfaces and equipment. However, it is a powerful oxidizer and must be handled with great care; breathing chlorine fumes or getting it on the skin can cause severe burns and even death. Always work with chlorine in a well-ventilated environment wearing protective clothing, especially for the hands, eyes, nose, and mouth.

A chlorine solution is a mixture of chlorine with water. If the mixture is too weak it will not be an effective disinfectant but if it is too strong it will be more dangerous to handle and may damage the surfaces being disinfected. In general, a 1 per cent or 2 per cent solution is appropriate for cleaning surfaces.

Chlorine is available in a variety of forms:

- Chlorine gas (100 per cent available chlorine) is commonly used at large water treatment plants but otherwise is not generally available. It is also dangerous to handle and not recommended for making cleaning solutions.
- Sodium dichloroisocyanurate (NaDCC) (56 per cent available chlorine) is currently the recommended product for chlorination in emergencies because of its ease of transport and storage, because it is safe to handle, and it has a long shelf life (three to five years). It is available in granule and tablet form.
- High test hypochlorite (HTH) (70 per cent available chlorine) is a granular powder usually supplied in drums.
- Sodium hypochlorite is a liquid commonly found in household disinfectant (5–15 per cent available chlorine) and laundry bleach (3–5 per cent available chlorine).
- Bleaching powder or chlorinated lime (about 30 per cent available chlorine when fresh).
- Chlorine tablets (60–70 per cent available chlorine) is used in swimming pools.

A 1 per cent chlorine solution will have 10 g of chlorine in a litre of clean water. The amount of chlorine product to be added therefore depends on the amount of available chlorine the chlorine source contains and also on the product's age. Most products should display the amount of available chlorine their product contains on the container; if they don't, contact the manufacturer. However, chlorine-based products are unstable and the chlorine will leach out over time. This is particularly the case with products such as bleaching power and household disinfectant. Always use fresh products when making up a solution.

Making a 1 per cent chlorine solution

- When mixing chlorine powders with water, always add the water to the chlorine, not the chlorine to the water.
- Set aside a fixed quantity of water depending on the total volume of solution you want to make.
- In another dry container, place the quantity of chlorine product you require. In emergencies, it may be difficult to determine the exact quantity of available chlorine to use; in these cases, the data in Table A2.1 will give an approximate guide to quantities.
- Add a small quantity of water to the chlorine and stir to a paste with a wooden spoon.
- Gradually add the remainder of the water, stirring all the time.
- Allow the solution to stand for a while so that the sediment can settle to the bottom.
- Draw off the clear liquid into a light-proof container with a sealable lid.
- Store in a cool dry place in the dark.
- The solution will lose its strength over time, so use it quickly.

Table A2.1 Preparation of 1 litre of 1 per cent chlorine solution

Chlorine source	Available chlorine (%)	Quantity required	Approx. measure
NaDCC	56	20 g	Depends on delivery method
HTH granules	70	14 g	1 heaped tablespoon
Bleaching powder	34	30 g	2 heaped tablespoons
Stabilized liquid bleach	25	40 g	2.5 tablespoons
Liquid household disinfectant	10	100 ml	7 tablespoons
Liquid laundry bleach	5	200 ml	14 tablespoons
Antiseptic solution	1	1 litre	No need to adjust

Notes: For other strengths of solution, multiply the quantities accordingly (i.e. for 2 per cent solution double the quantities shown). This does not apply to the antiseptic solution, which can only be made weaker.
This table delivers 1 per cent chlorine solution. To produce 1 per cent (ppm) chlorine in clean water, add 1 ml of the chlorine solution to 10 litres of clean water.
These figures are approximate; the final concentration of chlorine in the water body also depends on the quality of the water used for mixing and that of the receiving water body. Water with a high organic load or high suspended solids will have a lower chlorine concentration. If the strength of chlorine in the water body is critical, it should always be tested prior to use and the quantity of chlorine source material adjusted appropriately.

Source: (** Davis & Lambert, 2002).

APPENDIX 3

Excavation

Digging trenches and pits can be very dangerous if not done properly; it is one of the most common causes of accidents in construction. Any organization undertaking such work has a duty of care to its workers, whether employed directly or through a contractor, to ensure that they are appropriately protected and kept safe. Much of this appendix is taken from (** Davis & Lambert, 2002) but for more detailed advice consult the US Department of Labor (US Department of Labor, n.d.).

Stability of excavation

The angle at which the excavation can be dug depends on a number of factors:

- type and condition of soil;
- presence of groundwater;
- presence of surface water and rainfall runoff;
- loadings imposed by the spoil heaps, local buildings, traffic, etc.;
- depth of the excavation;
- time for which the excavation is left open;
- the weather.

Always protect excavations from rainfall and surface water runoff by constructing ditches uphill of the excavation to divert the water.

A rough idea of the appropriate excavation slope can be obtained from looking at the natural slope angles of exposed soils and at other construction slope angles in the area. Note that slope angles in dry soils are generally steeper than in wet soil. A guide to safe temporary slopes is given in Table A3.1.

Table A3.1 Temporary slope angles in different ground conditions

Ground type	Dry condition slope^	Wet condition slope^
Boulders	35–45	30–40
Gravel/sand	30–40	10–30
Silt	20–40	5–20
Clay	30–40	10–20

Notes: ^Slope measured in degrees from the horizontal.
       For more information on soil identification, see Figure A1.1.
Source: (** Davis & Lambert, 2002).

Temporary trench supports

Temporary support to trenches may be needed in unstable ground and where trenches are to be left exposed for some length of time (e.g. for deep trench toilets). Timber boards, bamboo, matted cane, and bush poles can be used to support the vertical sides of trenches temporarily. Struts and waling will be necessary as props between the walls, as shown in Figure A3.1. Struts should be placed at least every 1.8 m but may need to be closer, depending on the trench use. For example, in trench toilets, struts should be placed in line with each cubicle partition to prevent fouling. A second-level strut will be required if the depth of the trench exceeds 2 m.

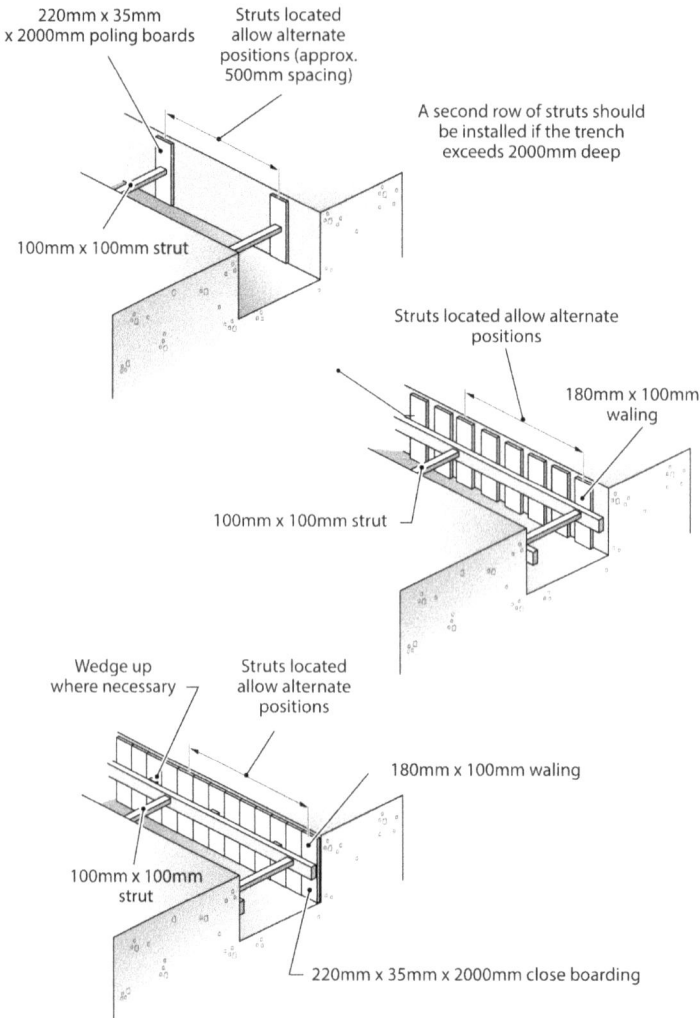

Figure A3.1 Timber support systems for deep trench toilets

APPENDIX 4

Emergency sewer networks

Sewer design has two components:

1. Short sewers connected to individual or small groups of fittings and delivering to a larger sewer or a local outfall
2. Large sewers carrying sewage from a group of connector sewers, as shown in Figure A4.4.

Designing short connector sewers

The minimum size of sewer pipe connected to an appliance is normally controlled by the diameter of the appliance outlet pipe. Table A4.1 recommends minimum figures for general use but the actual size used should never be less than the diameter of the appliance outlet pipe. The branch length refers to the maximum distance from the first appliance to the discharge to a main sewer pipe. The term 'unvented' indicates that the pipe does not have its own ventilation pipe. If the pipe has its own ventilation shaft, the number of appliances and the length of the pipe can be extended.

In general, in emergencies where water is scarce, the smaller the pipe diameter and the flatter the gradient the better. The main mechanism transporting solids is hydrostatic pressure building up behind the solids, pushing them along. Large-diameter pipes at steep gradients quickly lose any retained liquids, allowing the solids to dry out on the pipe bed and making them difficult to move (Figure A4.1). If the ground slope makes shallow sewer gradients impossible, consider laying short lengths of shallow gradient pipe, each followed by a drop manhole.

Table A4.1 Common appliance discharge pipes (unvented)

Appliance	Max. no. to be connected	Max. length of branch pipe (m)	Min. pipe size (mm)	Gradient (mm fall per m run)
WC outlet > 80mm	8	15	100	18^^–90
WC outlet < 80 mm	1	15	75^^^	18–90
Urinal – bowl		3^	50	
Urinal – trough		3^	65	18–90
Urinal – slab^^^^		3^		
Washbasin or bidet	3	1.7	30	18–22
		1.1	30	18–44

(Continued)

Table A4.1 Continued

Appliance	Max. no. to be connected	Max. length of branch pipe (m)	Min. pipe size (mm)	Gradient (mm fall per m run)
		0.7	30	18–87
		3.0	40	18–44
	4	4.0	50	18–44

Notes: ^Should be as short as possible to prevent deposits.
^^May be reduced to 9 mm on long drain runs where space is restricted but only if more than one flushing toilet is connected.
^^^Not recommended where sanitary towels are disposed in flushing toilets because of blockages.
^^^^Slab urinals longer than seven persons should have more than one outlet.
Source: (Ministry of Housing, Communities & Local Government (UK), 2010)

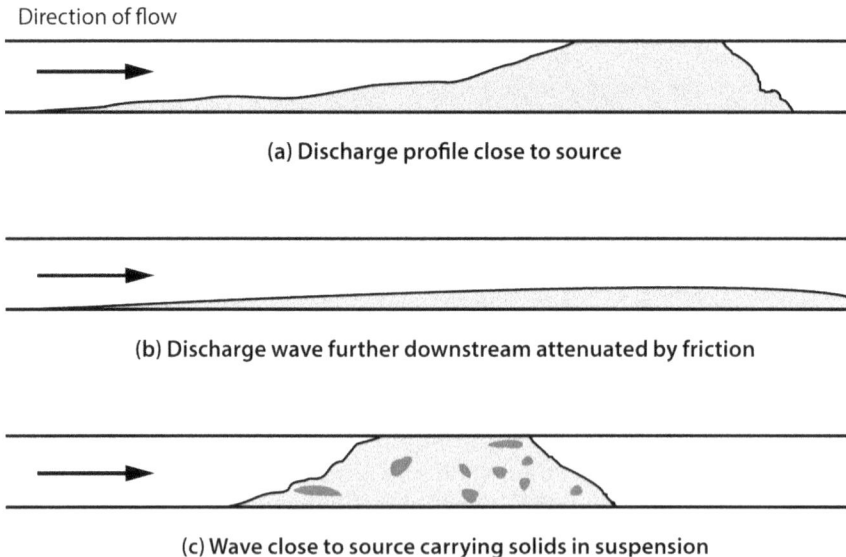

Direction of flow

(a) Discharge profile close to source

(b) Discharge wave further downstream attenuated by friction

(c) Wave close to source carrying solids in suspension

Figure A4.1 Solids transport mechanisms in the upper reaches of a sewer network

Designing main sewer pipes

Sewers should have the capacity, when flowing 75 per cent full, to carry the peak flow from all the appliances connected upstream of the pipe being considered. The total daily wastewater flow can be estimated by totalling all the appliances connected to the sewer and using the design figures for per capita usage and numbers of users per appliance given in Table 4.2. Not all the water used by appliances will end up in the sewer; some will be 'lost' by being either used for other purposes or spilt. The amount of water lost directly from toilet appliances is usually very low, but water not entering the waste system from a tap stand will be considerable (as most is supposed to be carried away by users). Estimating the proportion of water delivered to a community that ends up in the sewer has to be a judgement. In normal domestic environments a figure of about 85 per cent is used, but it could be very different in a refugee camp.

Not all appliances will be in use at the same time and there will be times when there is very little wastewater flow at all. This leads to peaks in the flow that are greater than the average flow for the day. Normal practice is to design on the basis of a multiple of the average day flow such that:

Peak flow (litres/hr) = k_1. × k_2. × total daily water consumption (litres)/24

Where k_1 = percentage of water supplied entering the sewer network

k_2 = peak factor – normally in the range 2 to 4, but for very low flows such as those found in small sewer networks it can be much higher, as shown in Figure A4.2.

The size of sewer pipe is a function of the peak flow rate and the sewer gradient, as shown in Figure A4.3 and Table A4.2. An example for a temporary settlement is shown in Box A4.1.

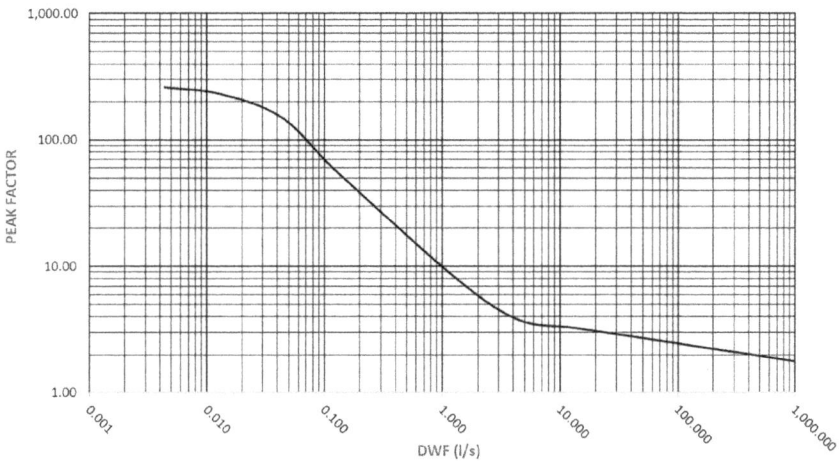

Figure A4.2 Peak factors for very low flows
Source: Adapted from (Haws, n.d.)

Table A4.2 Recommended minimum gradients for foul sewers

Peak flow (litres/second)	Pipe size (mm)	Minimum gradient (1 in …)	Maximum capacity (litres/second)
< 1	75	40	4.1
	100	40	9.2
> 1	75	80	2.8
	100	80^	6.3
	150	150^^	15.0

Notes: ^Minimum 1 flush toilet.
         ^^Minimum 5 flush toilets.
Source: (Ministry of Housing, Communities & Local Government (UK), 2010)

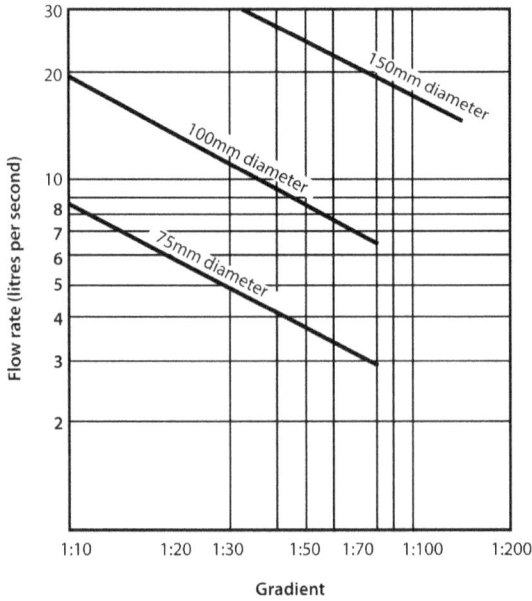

Figure A4.3 Discharge capacities of foul sewers running 75 per cent full
Source: Adapted from (Ministry of Housing, Communities & Local Government (UK), 2010)

Box A4.1 Example of small sewer network design

Figure A4.4 shows a proposed new temporary settlement for 96 families contained in eight community blocks with a central communal area. Design a sewer network to serve the toilet blocks.

Each community covers an area of 4,800 m² and consists of 12 family plots, each one 20 m × 20 m (400 m²). There is a shared toilet block for four plots including water tap, toilet (pour flush), bathing, and laundry areas.

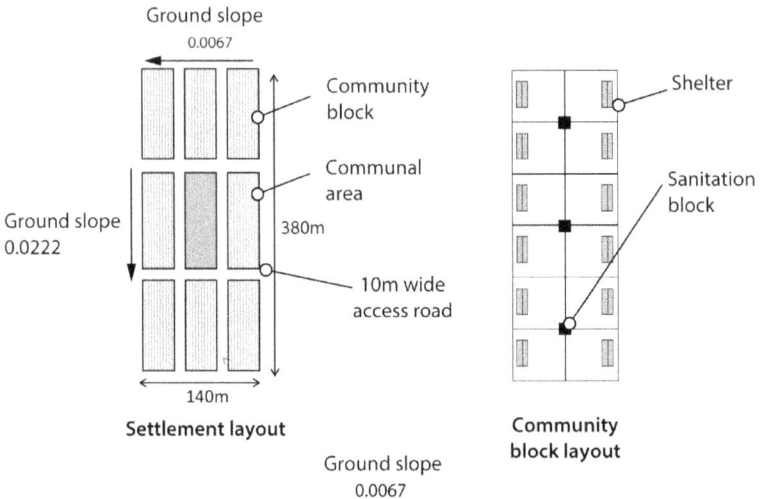

Ground slope 0.0067

Community block
Communal area
10m wide access road

Ground slope 0.0222

380m
140m

Settlement layout

Ground slope 0.0067

Shelter
Sanitation block

Community block layout

(Continued)

Box A4.1 Continued

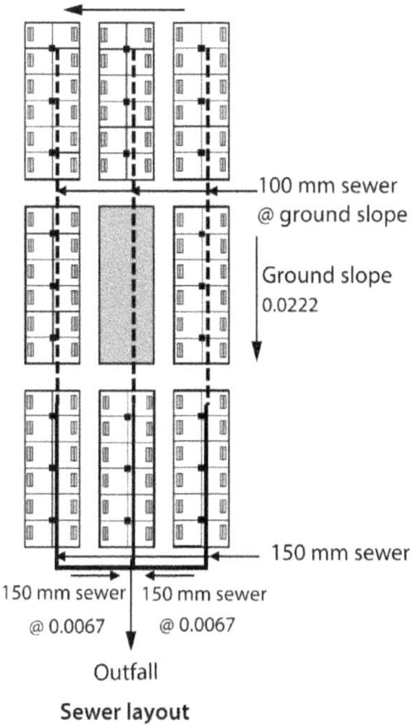

100 mm sewer
@ ground slope

Ground slope
0.0222

150 mm sewer

150 mm sewer | 150 mm sewer
@ 0.0067 | @ 0.0067

Outfall

Sewer layout

Figure A4.4 Case study block and sewer layout

Assumptions		Wastewater flow	
Total population	500	Water used per plot (litres/day)	130
Number of housing plots	96	Sewage generated per plot (litres/day)	104
Average family size	5.2	25% leakage per plot (litres/day)	6
Per capita water consumption (litres/day)	25	Total per plot including leakage (litres/day)	110
Percentage entering sewer system	80%	Sewage generated per 4 family toilet blocks (litres/day), known as dry weather flow (DWF)	442
Water point leakage (% of supply)	25%	Sewage generated per 12 family community blocks (litres/day)	1,325
		Total sewage generated (litres/day)	10,600

Sewer design
Assume the sewer layout shown in Figure A4.4.

Sewer unit	DWF (l/d)	DWF (l/s)	Peak factor^	Peak flow (l/s) ^^	Min. gradient ^^^	Ground slope ^^^^	Sewer gradient ^^^^^	Pipe size (mm) ^^^^^^
1 × family toilet block	442	0.005	260	1.33	0.0125	0.0222	0.0222	100

(Continued)

Box A4.1 Continued								
2 × family toilet blocks	883	0.010	240	2.45	0.0125	0.0222	0.0222	100
1 × community block	1,325	0.015	230	3.53	0.0125	0.0222	0.0222	100
2 × community blocks	2,650	0.031	190	5.83	0.0125	0.0222	0.0222	100
3 × community blocks	3,975	0.046	160	7.36	0.0067	0.0222	0.0222	150
Block collector sewer	3,975	0.046	160	7.36	0.0067	0.0010	0.0067	150

Notes: ^Taken from Figure A4.2 using DWF.
   ^^Multiplying DWF (litres/second) by peak factor.
   ^^^From Table A4.2. Check peak flow to decide whether to use figures for <1 or >1 peak flow, select smallest pipe diameter (assuming 100 mm is the smallest allowable size) that carries more than the peak flow, and read off the minimum gradient (expressed as a decimal). If the flow is greater than the maximum allowed for the largest pipe, use Figure A4.3 to select a different pipe size or sewer gradient.
   ^^^^The measured ground slope in the direction of sewer travel. In this case it is given in Figure A4.4.
   ^^^^^The greater of the minimum gradient and ground slope.
   ^^^^^^The pipe size selected.

Depth of cover

The minimum depth at which a sewer is laid depends on a variety of factors but mainly on the type of loading that is likely to be imposed on it from above and the depth of soil freezing in cold climates. Table A4.3 suggests minimum depths for different superimposed loads unless the maximum depth of soil freezing is greater. Appendix 3 contains details on safe trench excavation.

Pipe bedding

A common cause of sewer failure is the improper support to the pipe provided during laying. Figure A4.5 shows the recommended bedding method for thermoplastic pipes.

Table A4.3 Limit of cover for thermoplastic pipes

Nominal size	Laid in fields	Laid in light traffic roads	Laid in heavy traffic roads
100–300mm	0.6–7.0 m	0.9–7.0 m	0.9–7.0 m

Note: Sewers must always be laid below the depth of water pipes to prevent cross-contamination.
Source: (Ministry of Housing, Communities & Local Government (UK), 2010).

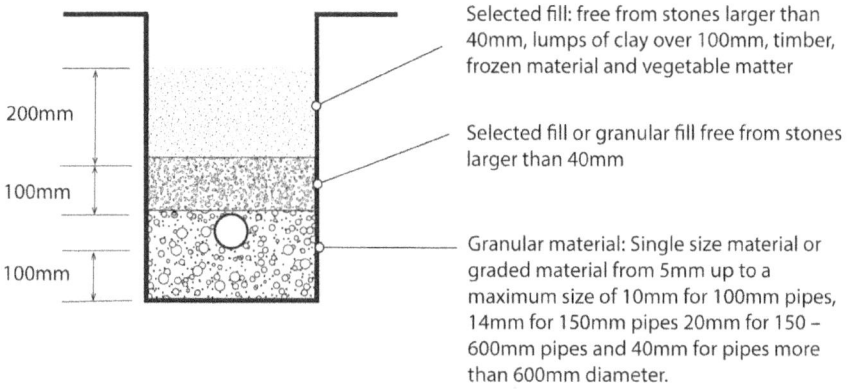

Selected fill: free from stones larger than 40mm, lumps of clay over 100mm, timber, frozen material and vegetable matter

Selected fill or granular fill free from stones larger than 40mm

Granular material: Single size material or graded material from 5mm up to a maximum size of 10mm for 100mm pipes, 14mm for 150mm pipes 20mm for 150 – 600mm pipes and 40mm for pipes more than 600mm diameter.

200mm

100mm

100mm

Figure A4.5 Recommended method for pipe bedding
Source: Adapted from (Ministry of Housing, Communities & Local Government (UK), 2010)

APPENDIX 5

Wastewater settlement test

The settlement test is an analysis of the settling characteristics of any wastewater or sludge containing suspended solids. It is simple and can be undertaken at the treatment plant by relatively unskilled workers to feed into plant design or monitoring. The test requires four basic pieces of equipment:

- A clear container to hold the wastewater: A 1.5 litre glass or plastic beaker with graduation marks on the side is the best (such as is shown in Figure A5.1), but if that is not available any clear container can be used with graduation marks added manually. For emergency purposes, the graduation marks do not need to be as detailed as shown in the figure.
- A timer or clock to track elapsed time.
- A paddle or other mixing device: This can be a clean straight wooden, glass, or plastic stick about 10 mm in diameter and longer than the depth of the container.
- A place to record the readings.

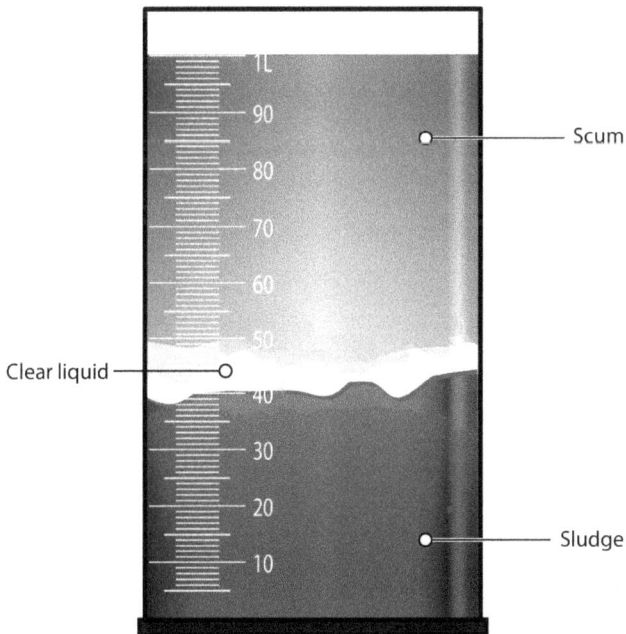

Figure A5.1 Graduated beaker for settlement test
Source: Adapted from (Treatment Plant Operator, n.d.)

The test is carried out as follows:

- In a container with a sealable lid, collect at least 2 litres of fresh, well-mixed sewage or faecal sludge after it has passed through the coarse filter. In some cases, the settlement process takes place at later stages in the treatment process, in which case collect the sample from the last point in the biological treatment process before the settlement tank you are designing or testing.
- Mix the contents of the sample container by inverting several times. Do not violently shake or agitate the sample.
- Gently fill the test container to the 1 litre mark. Run the test on a flat surface, out of direct sunlight. (Direct sunlight warms one side, causing uneven settling.)
- Gently mix the contents with the stirrer one more time, slowing the liquid current almost to a standstill.
- Start the timer or note the start time.
- Every five minutes, measure and record the depth of the settled sludge and the thickness of any floating scum. Sometimes there are high and low spots; use an average in these instances. Also note the colour and consistency of the liquid between the sludge and scum.
- After 30 minutes, increase the period between measurement to 10 minutes, and after 1 hour to 15 minutes.
- Once the test is complete, use graph paper or a computer program to enter the data into a trend chart such as the one shown in Figure A5.2.
- If undertaking the test for design purposes, repeat it at various times of the day and night for as long as is practical.

The test results in Figure A5.2 show that most of the settlement has taken place in 60 minutes, but for design purposes a retention time of 90 to 120 minutes would be suggested because of disturbances to the flow within a large tank.

SETTLEABILITY TEST (Example)

Location: _____   Date: _____

Start time: _____   Sample volume: 1.0 litres _____

Sludge liquid depth: 200 mm _____

Time (mins)	Sludge thickness (mm)	Scum thickness (mm)	Liquid consistency
5	0	0	
10	12	5	
15	23	9	Some stratification noted
20	32	12	
25	39	14	Mainly clear of solids
30	44	15	
40	50	15	Grey coloured and opaque
50	53	16	
60	54	16	No change in colour
70	54	16	
90	55	16	
105	55	16	
120	56	16	

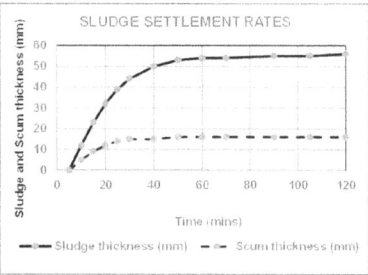

Figure A5.2 Specimen settlement test data

Settlement tests should also be carried out regularly once the unit is operational. Changes in settlement times could indicate changes in the content of the incoming waste. Tests should also be carried out on the unit effluent as these will highlight problems with the settlement process.

The test can also be used to design chemical treatment processes such as the use of calcium hydroxide for sludge dewatering (see Chapter 9, section 7).

APPENDIX 6

Tank flotation

A watertight tank constructed underground may be perfectly stable while the soil is dry but if the groundwater level rises the tank may start to float (Figure A6.1).To prevent this happening, the weight of the tank has to be heavier than the weight of the water taking up the same volume as the tank. Box A6.1 gives a worked example.

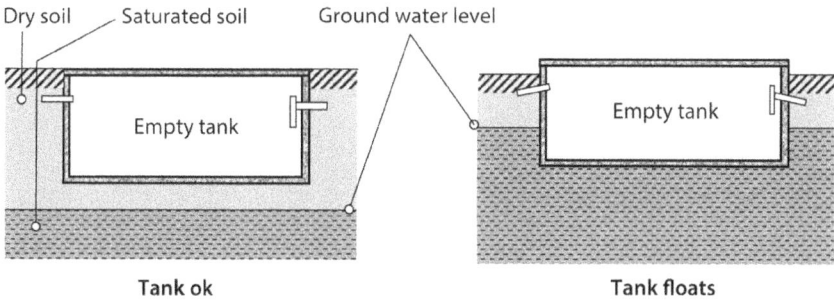

Figure A6.1 Effects of flotation

Box A6.1 Example test for text flotation

Using the design example given in Box 9.8:
Assume the thickness of the tank base is 150 mm and the tank walls and roof are 100 mm.
Assume the tank is constructed mainly of concrete.
Total dimensions of the tank are: 2.2 + 1.1 + 0.2 = 3.4 m long
1.1 + 0.2 = 1.3 m wide
1.8 + 0.1 + 0.15 = 2.05 m deep
Total volume is 3.4 × 1.3 × 2.05 = 9.061 m³ = volume of water that would be displaced if the groundwater level reached ground level.
Weight of water displaced = 9.061 × 1,000 (density of water) = **9,061 kg**
To calculate the total weight of the empty tank, find the volume of the materials used to construct it and multiply by their density.
Volume: Roof 3.4 × 1.3 × 0.1 = 0.442 m³
Side walls 1.3 × 1.8 × 0.1 × 3 (inc. dividing wall) = 0.702 m³
Base 3.4 × 1.3 × 0.15 = 0.663 m³
Total volume 1.807 m³
Weight of tank 1.807 × 2,240 (density of concrete) = **4,048 kg**
Therefore, the tank would float if the water table reached ground level. However, if the water table were likely to reach a lower level, then the volume of the tank submerged would be less and the tank may not float.
Remember: this is only a problem when the tank is empty. The weight of the wastewater in the tank will add to the weight of the tank and prevent floatation. That is why it is always good practice to fill the tank with water after desludging.

APPENDIX 7
Additional treatment options

Infiltration mounds

If the subsoil is impermeable (clay or rock) or the water table is close to the surface, the infiltration system can be placed in a mound, as shown in Figure 10.6. A mound of porous sand has, near its top, a gravel bed containing a network of porous distribution pipes connected via a pumping and dosing chamber to the effluent source, for example a septic tank. The original soil surface is disturbed (e.g. by ploughing) to break up its structure and improve permeability. The distance to the water table or bedrock governs the height of the mound. The plan area of the mound is decided by the permeability of the natural topsoil. The area must be large enough to infiltrate the total volume of effluent produced.

Mound systems are designed to treat domestic wastewater. Stronger wastes such as faecal sludge effluent must be diluted to a similar strength as domestic waste. The primary purpose of a mound is to filter fine particles from the effluent and disperse it over a wider area so that it can be more readily absorbed by the ground below. However, there is some evidence that considerable effluent purification also takes place. The design of infiltration mounds varies slightly depending on the problem. Where there is a high water table, the mound should be constructed on level ground; where the problem is impermeable soil, the mound should be constructed on a slightly sloping site (ideally 0–12 per cent). Unlike infiltration trenches, infiltration takes place primarily through the base of the gravel bed surrounding the distribution pipes and is therefore a function of the material used for constructing the mound. A graded gravel pack (layers of media of gradually decreasing size, usually around 100 mm thick) is necessary between the gravel bed and the sand and between the sand and the subsoil to prevent the two media combining. Alternatively, the gravel packs can be replaced by porous geotextile fabrics. Mounds are aerated through ventilation pipes connected to the gravel bed and soil/sand interface to prevent the system becoming anaerobic. Other key design parameters are given in Table A7.1 and a case study is shown in Box A7.1.

Table A7.1 Design data for infiltration mounds

Data	Value	Comments
Gravel trench width	5.0–6.5 m recommended	
Gravel trench length	4.0–35.0 m	Depends on the flow rate and the infiltration capacity of the sand in the mound (Table 4.3).
Gravel trench depth	Minimum 0.2 m	0.15 m below the distribution pipe + pipe diameter + cover.
Filter bed depth (below the distribution pipe gravel layer)	Not less than 1.2 m to the water table or top of the impermeable layer	Depends on the infiltration capacity of the original topsoil.
Minimum depth of water table below original ground level	0.5 m	
Minimum depth below ground level to fully impermeable layer	0.3 m	
Infiltration capacity of soil surface	Approx. 10 litres/m²/day	Topsoil generally has a higher permeability than the ground below because it has been disturbed and has a more open texture.
		If possible, measure the permeability; when not possible, this number is suggested.
Linear loading rate	Less than 0.125 m³/m	Determined by dividing the design flow by the length of the gravel bed.
Water dilution ratio for faecal sludge	Between 4:1 and 20:1	Depends on the strength of the inflow.
		Effluent from fresh faecal sludge is nearer the lower end of the range and septage from infrequently emptied pits and public toilets will be nearer the upper end.
		Alternatively, the infiltration rates can be reduced by similar ratios.
Porous pipe diameter	25–38 mm	When pumping, small pipes makes it easier to obtain uniform distribution.
Pipe slope	Horizontal	This will mean that the pipes nearest the inlet receive effluent first, but designing and accurately constructing pipes with a very shallow slope is very difficult.
Filter dosing frequency	2–4 times a day	To allow time for the filter to rest between dosings and remain aerobic.

(Continued)

Table A7.1 Continued

Data	Value	Comments
Bedding gravel	Around 10–20 mm diameter	Gravel should be washed to remove fine particles.
Gravel bed dimensions	13–32 m long Less than 2.8 m wide	Beds can be bigger than this but it is not recommended.
Sand characteristics	Less than 20 per cent grain size greater than 2.0 mm Less than 5 per cent finer than 0.053 mm	
Maximum mound side slope	33%	

Source: (State of Maryland Dept. of the Environment, 2016 February), (Phelps, 2005)

Advantages
- Provides an infiltration system for areas with impermeable soils close to the surface or a high water table.
- The soil utilized is close to the surface, which is typically the most permeable.
- No direct discharge into surface waters.
- Works in most climates, except frozen soil.
- Uniform distribution of effluent over the infiltration pipe network makes it more efficient and less likely to clog.

Disadvantages
- Unsuitable for large wastewater volumes.
- More expensive than below-ground infiltration systems.
- Requires pumps and dosing syphons.
- Only recommended for fresh domestic sewage with low BOD levels (150–250 mg/litre).
- System takes a few weeks for the purifying biofilms to become fully effective.
- Requires regular monitoring and maintenance.

Sand filters

A single-pass sand filter system (Figure 10.7.) is a secondary treatment after suspended solids have been removed. It filters the influent through sand before sending it for final treatment or direct discharge (Figure A7.2). Other similar filters use peat, pea gravel, crushed glass, or other experimental media, but sand is the best understood and the most predictable. Treatment mechanisms in a sand filter include physical filtering of solids, ion exchange (alteration of compounds by binding and releasing their components), and decomposition of organic waste by soil-dwelling bacteria. Although sand filters can be used as a final treatment process, they are most commonly used as an intermediate stage to reduce the organic loading on subsurface infiltration systems in low permeability soils or infiltration mounds. A properly operating sand filter receiving domestic-strength influent should produce high-quality effluent with less than 10 mg/litre BOD, less than 10 mg/litre TSS and less than 200 cfu/100 ml (Gustafson, et al., 2001).

Partially treated wastewater flows into a sump or lift tank (similar to Figure A7.1). A pump introduces the influent at the top of the watertight sand filter, using pressure distribution to apply the wastewater evenly to the filter surface to maximize treatment. A timer is used to dose the entire surface of the filter intermittently with wastewater. This draws oxygen from the atmosphere through the sand medium and its attached microbial community. Sand filter systems may also be successfully retrofitted into drainage fields that have failed because of excessive organic loading or from lack of maintenance. Sand filters are commonly used for treating effluent from small communities and institutions.

Box A7.1 Calculating the size of an infiltration mound

Case study

Design an infiltration mound to dispose of the effluent from the septic tank designed in Box 9.8.
Assume the soil is clay loam that has been ploughed and the water table is at a depth of 0.5 m but is highly saline and unsuitable for a drinking water supply.
Daily effluent flow rate = 2.0 m³
Infiltration rate for coarse sand = 50 litres/m²/day (Table 4.3)
Infiltration area into the top of the sand mound = 2.0 × 1000/50 = 40 m²
However, the effluent is strong, undiluted faecal waste. Therefore, either dilute the effluent by a factor of 4 or increase the filter area by a factor of 4. Suggest the latter as it is easier to manage if land is available.
Design infiltration area = 40 × 4 = **160 m²**
Calculate the shape of the gravel bed below the infiltration pipes
Assume a maximum width of gravel bed of 6.5 m
Therefore, the length of the gravel bed = 160/6.5 = **24.6 m**
This is a length to breadth ratio of 3.8:1
Surface level infiltration area
Soil surface infiltration rate = 10 litres/m²/day
Soil infiltration area = 2.0 × 1,000/10 = 200 m²
Again, we must increase the surface area by a factor of 4 to allow for the strength of the effluent
Design area at ground level = 200 × 4 = **800 m²**
Using the same length to breadth ratio as the gravel bed gives width of the bed as (800/3.8)^(1/2) = **14.5 m**
Total length of the base of the mound = 14.5 × 3.8 = **55.1 m**

(Continued)

Box A7.1 Continued

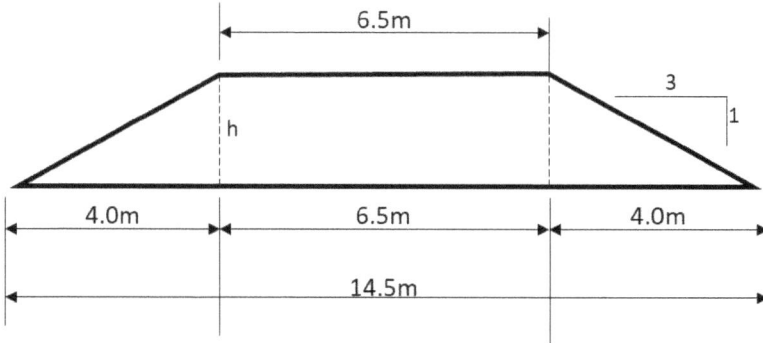

The figure above shows a cross-section through the filter bed below the gravel bed.
The height of the filter bed (h) = 4/3 = **1.33 m**
This is greater than the minimum depth shown in Table A7.1 so is OK.
Remember: the mound will be higher and wider than this because of the topsoil surrounding the filter.

Source: Adapted from (State of Maryland Dept. of the Environment, 2016 February).

Single pass sand filters can be above ground or buried, although ready access is required to the filter top for routine maintenance. The entire filter unit is contained in an impermeable liner. Underdrain pipes and a graded layer of washed gravel or crushed rock are placed at the bottom of the filter bed, with the finer gravel on top of the coarser gravel to keep the media grains from washing into the underdrains. The filter media is then placed on top of the layer of fine gravel. Another graded layer of gravel is placed on top of the media bed and surrounds a network of distribution pipes. However, the order is reversed this time and the finer gravel is placed under the coarser gravel closest to the media bed. A geotextile fabric is placed on top of the entire filter bed to prevent contamination. Other design criteria are given in Table A7.2. A design example is shown in Box A7.2. It is useful if the top gravel layer is removable to simplify cleaning the top of the sand. This may just mean raking the top of the sand bed but could include removing a thin layer of blocked sand. Dividing the bed into two units allows the filter to keep working during cleaning/maintenance.

Advantages
- Can provide intermediate-quality effluent usually requiring further treatment.
- Enables development in difficult sites.
- Can remedy an existing malfunctioning system.
- Does not pollute surrounding ground.
- Can be fabricated off-site for multiple applications.

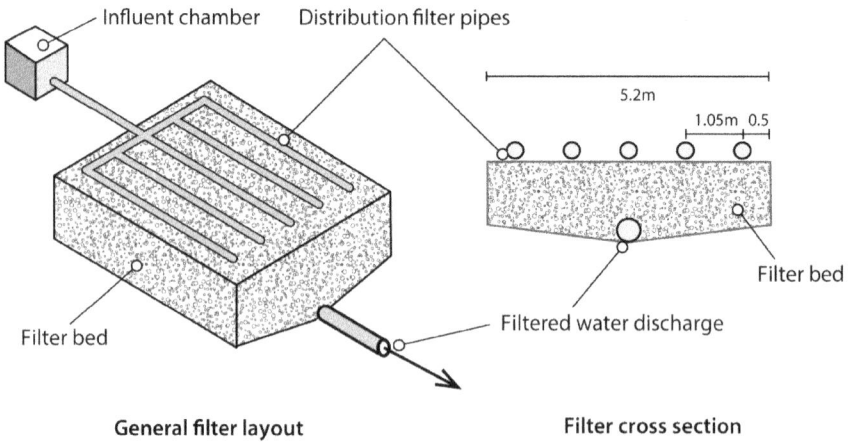

General filter layout   **Filter cross section**

Figure A7.1 Sand filter layout

Disadvantages
- More expensive than below-ground infiltration systems.
- Requires pumps and dosing syphons.
- Only suitable for fresh domestic sewage with low BOD levels (150–250 mg/litre).
- System takes a few weeks for the purifying biofilms to become fully effective.
- Regular maintenance essential.

Table A7.2 Design data for sand filters

Data	Value	Comments
Maximum flow rate	450 m³/m²/day	Domestic strength sewage.
		Dilute waste from stronger influent such as faecal sludge (4 times dilution suggested but retain maximum flow rate).
Filter sand depth	0.6–1.0 m washed	
Filter bed cleaning	2–6 months	Depends on the dosage strength and rate but generally longer in cold climates than in hot.
Minimum working temperature	Greater than 5°C	Unsuitable for long periods below 0°C.
Filter dosing rate	0.08–0.2 m³/m²/day	Dosage rate increases with ambient temperature.
Dosing frequency	2–4 times a day	To allow time for the filter to rest between dosings and remain aerobic.
Distribution pipe diameter	25–38 mm	When pumping, small pipes make it easier to obtain uniform distribution.

(Continued)

Table A7.2 Continued

Data	Value	Comments
Filter sand size	0.3–3 mm washed	
Bedding gravel	10–20 mm washed	Around the distribution and collection pipes.
Intermediate gravel	3–10 mm washed	Between bedding gravel and filter sand.
Gravel bed thickness	Approx. 0.25 m	
Bed shape	Not critical but usually rectangular	
Distribution pipe spacing	0.6–1.5 m apart	

Source: Adapted from (Gustafson, et al., 2001), (Taylor, 1997)

Box A7.2 Calculating the size of a sand filter bed

Case study
Design a sand filter to dispose of the effluent from the septic tank designed in Box 9.8.
Daily effluent flow rate = 2.0 m^3

However, the effluent is strong, lightly diluted faecal waste. Therefore, either dilute the effluent by a factor of 4 or increase the filter area by a factor of 4. Suggest the former as it is easier to distribute around the filter bed.
Design flow = 2 × 4 = **8 m^3** – below maximum flow so OK
Assume a filter dosing rate of 0.1 m^3/m^2/day (Table A7.2)
Infiltration area into the top of the sand mound = 8.0/0.1 = 80 m^2
Assume the shape in plan is rectangular in ratio of 3:1
Width = (80/3)$^{1/2}$ = 5.2 m
Length = 15.6 m
Adjust number of pipes so that distance between pipes lies within the range suggested in Table A7.2.
Try 5 pipes = [5.2–(0.5 × 2)]/4 = 1.05 m – this is OK
Assume sand depth of 1.0 m and top and bottom gravel bed depths of 0.25 m (Table A7.2).

Source: Adapted from. (Gustafson, et al., 2001), (Taylor, 1997)

APPENDIX 8
Glossary

Aerobic	Wastewater contains dissolved oxygen that micro-organisms can use as an energy source.
Anaerobic	In wastewater containing no oxygen, micro-organisms must create chemical reactions to produce oxygen.
Biological oxygen demand (BOD)	A measure of the oxygen demand exerted by the readily bio-oxidizable organic material contained in a wastewater sample over a given time period. BOD is normally determined over a five-day period at 20°C and is referred to as BOD5.
Blackwater	Wastewater mainly containing excreta, flush water, and anal cleaning materials.
$Ca(OH)_2$	Chemical formula for calcium hydroxide, commonly known as hydrated lime.
Colony forming unit (cfu)	A unit commonly used to estimate the concentration of micro-organisms in a test sample.
Commode	A chair or piece of furniture containing a waterproof container into which users may urinate or defecate.
Communal toilets	Blocks of multiple toilets shared by a group of households, sometimes segregated by gender and age. They are arranged for the exclusive use of targeted households (i.e. at the centre of a block) and are not intended for use by the general public.
Cubicle	Sometimes called the superstructure, this consists of the external elements of the toilet, including the walls, roof, floor, lockable door, lighting, and access.
E. coli	E. coli (short for Escherichia coliform) is a type of faecal coliform bacteria that is commonly found in the intestines of animals and humans. E. coli in water is a strong indicator of sewage or animal waste contamination. Sewage and animal waste can contain many types of disease-causing organisms. Consumption may result in severe illness; children under five years of age, those with compromised immune systems, and the elderly are particularly susceptible.

Effluent	Generally a liquid waste or sewage discharged into a river or the sea. However, in this book it also refers to the liquid leaving any element of a wastewater treatment plant.
Emergency phases	The emergency response measures change with time and the response is commonly divided broadly into phases. Unfortunately, the number and description of phases are not standardized. In this book, a three-phase description is used: acute response, stabilization, and recovery. A more detailed description of the phases is given in Box 1.1.
Excreta	Waste matter discharged from the body, primarily faeces and urine.
Faecal sludge	Faecal sludge is primarily human excreta, usually combined with other elements such as anal cleaning water, toilet paper, flushing water, soap, bathing water, and kitchen and laundry wastes. It may also contain varying quantities of silt and grit, grease, menstruation products, and domestic garbage. Sometimes the sludge is fresh (i.e. recently produced) and sometimes aged, such as when taken from a pit toilet or septic tank. Such wastes tend to be of higher concentration because of the higher suspended solids component and the partial biochemical breakdown of the organic elements in the waste.
Faecal sludge management	The deposition, storage, collection, transport, treatment, and safe end use or disposal of faecal sludge and liquids.
Garbage	Another name for solid waste, usually applied to general domestic waste.
Gender-based violence	A wide range of violence and abuse, committed primarily but not exclusively against women by men.
Global WASH Cluster	A group of organizations with expertise in WASH working together under the management of UNICEF to improve the quality and effectiveness of WASH services in emergency responses.
Grey water	Wastewater from bathrooms (not toilets), sinks, washing machines, and other kitchen appliances.
Household toilets	Toilets used by a single household.
Humanitarian programme cycle	This sets out the sequence of actions that should be taken to prepare for, plan, manage, deliver, and monitor collective humanitarian responses (e.g. human waste management).

Hydrated lime	See $Ca(OH)_2$.
Hydrogeological	The study of groundwater, where it is found and distributed, and its movement below the Earth's surface.
Inclusive design	The design of an environment so that it can be accessed and used by as many people as possible, regardless of age, gender, or disability. It applies to buildings surrounding open spaces and wherever people go about everyday activities. Inclusive design keeps the diversity and uniqueness of each individual in mind. To do this, built environment professionals should involve potential users at all stages of the design process, from the design brief and detailed design through to construction and completion. Where possible, it is important to involve disabled people in the design process (Inclusive Design Hub, 2022).
Influent	Wastewater or other liquid, untreated or partially treated, flowing into a treatment process or treatment plant.
Infrastructure	The basic physical structures and facilities (e.g. toilets, solid waste trucks, treatment plants) needed for the operation of a society or enterprise.
Liquefy	To make a solid material such as pit sludge flow like a liquid, usually by mixing it with water.
Loam	A soil with roughly equal proportions of sand, silt, and clay.
Menstrual hygiene management	The specific hygiene and health requirements of girls and women during menstruation, such as the knowledge, information, materials, and facilities needed to manage menstruation effectively and privately.
Pathogen	A bacterium, virus, or other micro-organism that can cause disease.
Permafrost	Ground that is permanently frozen, usually a leftover from the last ice age.
Potable	Water that is safe to drink or can be made safe to drink with available treatment processes.
Public toilets	Public toilets can be used by any member of the public. They are typically installed in public places such as markets or mosques or along main thoroughfares. There is generally less sense of 'ownership with public toilets compared with communal toilets.

Putrescible	Organic material likely to decompose.
Sanitation	Although commonly used when discussing human faecal management, it has no concise or agreed definition. In general, it refers to public health measures necessary for the control of disease.
Septage	Another commonly used term for faecal sludge but it is slightly different. Septage refers specifically to faecal sludge collected from primary treatment processes such as septic tanks or to settlement tank sludge. It may also include the sludge from pit toilets.
Service delivery	The delivery of a service (e.g. toilet construction, solid waste collection, sullage channel maintenance) by one group (e.g. government department, business, community group, individual) to another group. It is usually a service that the receiving group is unable or unwilling to provide for itself.
Service provider	A company, organization, government department, or agency that delivers a service (e.g. empties pits, collects solid waste, manages and maintains wastewater treatment plants).
Sewage	Black and grey water conveyed on or below the ground in circular pipes (called sewers) or in channels of other cross-section.
Sewerage	A network of pipes and/or channels conveying sewage.
Shared toilets	Single toilets that are shared between a maximum of four households or extended family households. Shared toilets are often used as an emergency response intervention instead of constructing communal toilets. Shared toilets require that specific households are willing to share the use and maintenance of the facilities.
Short-circuiting	A condition that occurs when water flows along a nearly direct pathway from the inlet to the outlet of a tank or basin, often resulting in shorter contact, reaction, or settling times in comparison with the calculated or presumed detention times.
Slab	A flat, usually round, square, or rectangular plate covering a hole in the ground. Most commonly made of reinforced concrete but sometimes wood and rarely plastic or ferro cement.
Soffit	The top of the inside of a pipe.
Soil active zone	The soil layer that alternates between being frozen and thawed depending on temperature.

Solid waste	For the purposes of this book, all primarily non-liquid wastes generated by human activity and those resulting from the disaster itself.
SPHERE	Humanitarian Charter and Minimum Standards in Humanitarian Response.
Stakeholder	Any single or group of individuals (including individuals in organizations) with an interest or concern in something another individual or organization wishes to do or is responsible for.
Substrate	The surface or material on or from which an organism lives, grows, or obtains its nourishment.
Sullage	Wastewater from household sinks, showers, and baths, but not waste liquid or excreta from toilets.
Suspended solids	The tiny particles that float around in wastewater and do not settle out easily over time. Most wastewater treatment systems try to reduce suspended solids to less than 30 mg/litre before discharge. They can be reduced by both settlement and biological treatment. Most treatment plants use both processes.
Sustainable	An intervention that meets the needs of all of the people it is intended to assist for the whole time over which it is expected to function.
Toilet	The space, in total, in which people defecate and urinate. Additionally, for women and girls it should be a safe and dignified place where they can attend to their menstrual hygiene practices. The toilet includes the cubicle, floor, handwashing facilities, and interface. The room may also include other facilities such as washing, bathing, and laundry facilities. The space may be designed for individual family, shared, communal, or public use. Confusingly, it is also commonly known as the superstructure or cabin.
Toilet environment	The area between the user's residence and the toilet.
Toilet interface	The structure that collects excreta from the user and transfers it to the next element in the excreta management chain.
Toilet pan	The object into which people defecate. It is called a pedestal when the user sits on it and a squatting pan when squatted over.
Utilidor	An above-ground insulated conduit used for general utility services (including sewers), especially in arctic climates.

Viscosity	The resistance of a fluid (liquid or gas) to a change in shape, or movement of neighbouring portions of the same fluid relative to one another. The higher the viscosity the thicker the fluid and the more resistance it has to flow (e.g. treacle has a higher viscosity than water).
Wastewater	Water that has been used in the home, in a business, or as part of an industrial process.

APPENDIX 9
Abbreviations

ABR	Anaerobic baffled reactor
BOD	Biological oxygen demand
cfu	Colony forming unit
EEM	Emergency excreta disposal
ET	Evapotranspiration
FS	Faecal sludge
FSM	Faecal sludge management
HWM	Human waste management
MHM	Menstrual hygiene management
NFI	Non-food item
O&M	Operation and maintenance
PPE	Personal protective equipment
SWM	Solid waste management
TSS	Total suspended solids
UASB	Up-flow anaerobic blanket reactor
UDT	Urine diversion toilet
VIP	Ventilated improved pit toilet
WASH	Water, Sanitation and Hygiene Promotion
WSP	Waste stabilization pond
VT	Vermifilter toilet

References

References prefixed by '**' are considered key publications and have been used extensively in this book.

** Davis, J. & Lambert, R., 2002. *Engineering in emergencies: A practical guide for relief workers*. 2nd ed. London: ITDG.
** Gensch, R., Jennings, A., Renggli, S. & Reymond, P., 2018. *Compendium of Sanitation Technologies in Emergencies*. 1st ed. Berlin: German WASH Network (GWN), Swiss Federal Institute of Aquatic Science and Technology (Eawag), Global WASH Cluster (GWC) and Sustainable Sanitation Alliance (SuSanA).
** Harvey, Baghri, S. and Reed, B 2020. *UNHCR WASH Manual: practical guidance for refugee settings,* Geneva: UNHCR.
** Harvey, P. A., 2007. *Excreta Disposal in Emergencies: A Field Manual*. s.l.: WEDC, Loughborough University, UK.
** Harvey, P. A., Baghri, S. & Reed, R. A., 2002. *Emergency Sanitation: Assessment and programme design*. s.l.:WEDC, Loughborough University, UK.
** Jones, H. & Reed, R., 2005. *Water and sanitation for disabled people and other vulnerable groups: Designing services to improve accessibility*. s.l.: WEDC, Loughborough University, UK.
** Sphere Association, 2018. *The SPHERE Handbook: Humanitarian Charter and Minimum Standards in Humanitarian Response,*. Fourth Edition ed. Rugby: Practical Action Publishing.
** Taylcr, K., 2018. *Faecal Sludge and Septage Treatment. A guide for low and middle income countries*. Rugby: Practical Action Publishing.
** UNEP/OCHA Envionment Unit, 2011. *Disaster waste management guidelines,* Geneva: Joint UNEP/OCHA Environment Unit.
** WEDC. Loughborough University, 2023. s.l.: s.n.
** WEDC, n.d. [Art] (WEDC, Loughborough University, UK).
Adam-Bradford, A., Tomkins, M., Perkins, C. and Hunt, S., 2017. *Transforming land, transforming lives: Greening innovation and urban agriculture in the context of forced displacement,* Lewes, UK: The Lemon Tree Trust.
Adams, J., 1999. *Managing water supply and sanitation in emergencies*. Oxford: Oxfam.
Barnard Health Care, 2022. *Composting Plant Open Windrow*. [Online] Available at: https://www.barnardhealth.us/anaerobic-degradation/coo.html
Beesley, J., 2010. *Emergency female urinal, Haiti*. [Art].
Bibby, K., Fischer, R. J., Casson, L. W., Stachler, E., Haas, C. N. and Munster, V. J., 2015. Persistence of Ebola Virus in Sterilized Wastewater. *Environmental Science & Technology Letters*, 2(9), pp. 245–249.
Bidgee, n.d. *Pan Evaporation*. [Online] Available at: https://commons.wikimedia.org/wiki/File:Evaporation_Pan.jpg [Accessed January 2021].

Canadian Centre for Occupational Health and Safety, n.d. *Pushing & Pulling - Handcarts.* [Online] Available at: https://www.ccohs.ca/oshanswers/ergonomics/push2.html

Cement Concrete & Aggregates Australia, n.d. *Concrete basics: A guide to concrete practice.* [Online] Available at: https://www.ccaa.com.au/imis_prod/documents/ConcreteBasics.pdf

Center for Food Security and Public Health, 2003. *Cholera.* [Online] Available at: https://www.nj.gov/agriculture/divisions/ah/diseases/cholera.html#:~:text=cholerae%20is%20excreted%20in%20the,for%201%20to%202%20hours.[Accessed 16 April 2023].

Ciba-Geigy Agrochemicals, 1986. *Guide to soil identification.* Cambridge, UK: s.n.

Converse, J. C. & Tyler, E. J., 1987. On-site wastewater treatment using Wisconsin Mounds on Difficult Sites. *Transactions of the ASAE.,* 30(2), pp. 362–398.

Coultas, M. and Iyer, R. with Myers, J., 2020. *Handwashing Compendium for Low Resource Settings: A Living Document.* [Online] Available at: https://www.ids.ac.uk/publications/handwashing-compendium-for-low-resource-settings-a-living-document/

Curtis, V. C. S., 2003. Effect of washing hands with soap on diarrhoea risk in the community: A systematic review. *Lancet Infectious Diseases,* 3(5), pp. 275–81.

Deep Green Permaculture, 2009. *Three Simple Soil Tests to Determine What Type of Soil You Have.* [Online] Available at: https://deepgreenpermaculture.com/2020/07/23/three-simple-soil-tests-to-determine-what-type-of-soil-you-have/

Demiriz, M., n.d. The CIB WO62 36th International symposium water supply & drainage for buildings. pp. 356–363.

EAWAG, Sandec, 2008. *Sandec Training Tool 1.0 – Module 5 - Faecal Sludge Management.* [Online] Available at: http://www.susana.org/images/documents/07-cap-dev/c-training-uni-courses/available-training-courses/sandec-tool/05_fsm/module5_final.ppt

ELHRA, 2019. *Surface Water Management in Humanitarian Contexts: Practical guidance on surface water management & drainage for field practitioners,* London: Elhra.

ELRHA, n.d. *FSM selection.* [Online] Available at: https://www.elrha.org/ [Accessed March 2023].

Erlenbach, D. L. P., 1998. *Planning Guide for On-Site Greywater/Wastewater Disposal Systems for Recreational and Administrative Sites Pare II.* Washington, D.C.: US Department of Agriculture.

Flintoff, F., 1976. *Managment of solid wastes in developing countries.* New Dehli, India: WHO.

Franceys, R., Pickford, P. J. and Reed, R., 1992. *A guide to the development of on-site sanitation.* Geneva: WHO.

Georges, M., Robbins, D. M., Ramsay, J. E. & Mbeguere, M., n.d. *Chapter 4 Methods and Means for Collection and Transport of Faecal Sludge.* [Online] Available at: https://www.eawag.ch/fileadmin/Domain1/Abteilungen/sandec/publikationen/EWM/Book/FSM_Ch04_lowres.pdf

Global Education Cluster, 2010. *Education Cluster Handbook.* [Online] Available at: https://reliefweb.int/sites/reliefweb.int/files/resources/Full_Report_2178.pdf

Global WASH Cluster, 2009. *Water, Sanitation and Hygiene (WASH) Cluster Coordination Handbook.* [Online] Available at: https://mail.google.com/mail/u/0/?shva=1&pli=1#label/Committees%2C+Conferences+%26+Papers%2FEIE+Book/CllgCHrgDCMkVbXfGssbbKXS WnwPljbCcS LxtJtwzRbJXfxhSCJxLDDvjSHwzGMdZdhMhzQCGCL?projector=1&messagePartId=0.3

GOAL, 2016. *Review of Manual Pit Emptying Equipment Currently in Use and Available in Freetown and Globally.* [Online] Available at: https://wedc-knowledge.lboro.ac.uk/resources/pubs/Desk_Study_of_MPE_Technologies_GOAL_Sierra_Leone.pdf

Gustafson, D. M., Anderson, J. l. & Christopherson, S. H., 2001. *Single-pass sand filters,* s.l.: s.n.

Haws, T., n.d. *Sanitary Sewer Peaking Factors for Very Low Flows—A Study.* [Online] Available at: http://www.hawsedc.com/peakfact.php [Accessed May 2021].

Homeseptic Private Drainage Solutions, n.d. *A Guide to Drainage Field Design and Size Calculation.* [Online] Available at: https://homeseptic.co.uk/drainage-field-size-calculation-and-design/

Inclusive Design Hub, 2022. *About Inclusive Design.* [Online] Available at: https://inclusivedesign.scot/what-is-inclusive-design/[Accessed 2022].

Initial, 2021. *Colour Coding Guide.* [Online] Available at: https://www.initial.co.uk/colour-coding-guide/

ISF-UTS, SNV, 2016. *A guide to septage transfer stations,* s.l.: Institute for Sustainable Futures, University of Technology Sydney.

Jubair, M., Morris, J. J. & Ali, A., 2012. Survival of Vibrio cholerae in Nutrient-Poor Environments Is Associated with a Novel "Persister" Phenotype. *PLoS ONE,* 7 (9)(e45187).

Kamel, H. A. F., n.d. *Primary Sedimentaton Tank.* [Online] Available at: https://feng.stafpu.bu.edu.eg/Civil%20Engineering/726/crs-5633/Primary%20Sedimentation%20tank.docx

Kayaga, S. & Cotton, A., 2019. The characteristics of solid waste. In: *Solid waste management - Distance learning module.* s.l.: s.n.

Kayombo, S. M. T. S. A. K. J. H. Y. L. N. J. S. E., n. d. *Waste stabilization ponds and constructed wetlands design manual,* s.l.: UNEP-IETC and the Danish International Development Agency.

Lawrence, A. R, Macdonald, D. M. J., Howard, A. G., Barrett, M. H., & Pedley, S., Ahmed, K. M. and Nalubega M., 2001. *Guidelines for assessing the risk of groundwater from on-site sanitation,* s.l.: British Geological Survey.

Leblanc, M. et al., 2019. *Improving Sanitation in Cold Regions: Catalog of Technical Options for Household Level Sanitation,* s.l.: World Bank.

McBride, A. M. C. G. D. P. J., 2018. *Urine Diversion Dry Toilets: Standard Operating Procedures for Refugee Camps.* [Online] Available at: https://www.humanitarianlibrary.org/resource/urine-diversion-dry-toilets-standard-operating-procedures

McCartan, S., Stubbs, E., Crea, N., Kennard, A. and Blaikie, S., 2015. *Medical Support Vessel for 2025.* London, s.n.

Médecins Sans Frontières, n.d. *Medical Activities - Cholera.* [Online] Available at: https://img.msf.org/AssetLink/lpok3rx501airu03i533t0y753bofu7f.jpg

Medicins Sans Frontieres, 2010. *Public Health Engineerin in Precarious Situations.* 2nd ed. s.l.: s.n.

Ministry of Housing, Communities & Local Government (UK), 2010. *Approved Document H: Drainage and waste disposal.* [Online] Available at: https://www.gov.uk/government/publications/drainage-and-waste-disposal-approved-document-h

Mishra, G., n.d. *Evaporation and its Methods of Measurement.* [Online] Available at: https://theconstructor.org/water-resources/evaporation-and-its-measurement/4575

Mooijman, A., 2012. *Water, Sanitation and Hygiene (WASH) in Schools.* New York: UNICEF - Education section, Program division.

Morris, H., 2021. Global average per capita tissue consumption stands at above 5Kg - but 10Kg is possible. *Tissue World Magazine,* 26 March.

Noortgate, J. Van Den and Maes, P., 2010. *Public Health Engineering in Precarious Situations.* Paris: Medecins Sans Frontieres.

O`Riorden, M., 2009. *Investigation of Methods of Pit Latrine Emptying: WRC Project 1745: Management of sludge accumulation in VIP latrines,* s.l.: Partners in Development (Pty) Ltd.

OCTOPUS, n.d. *Crises.* [Online] Available at: https://octopus.solidarites.org/ [Accessed 28 October 2020].

Oxfam GB, 2018. *Sani tweaks. Best practices in sanitation,* s.l.: s.n.

Oxfam WASH, n.d. *Exit from WASH programmes.* [Online] Available at: https://www.oxfamwash.org/en/running-programmes/exit

Oxfam WASH, n.d. *Tiger Worm Toilet Manual.* [Online] Available at: https://www.oxfamwash.org/en/sanitation/tiger-worm-toilets

Oxfam, 2008. *Composting of Organic Materials and Recycling.* [Online] Available at: https://www.oxfamwash.org/sanitation/solid-waste/TB16%20Composting%20Organic%20Materials%20Recycling.pdf

Oxford Devises Ltd, 2023. *Whiz Freedom personal femail urinal,* s.l.: s.n.

Panafrican Emergency Training Centre, 2002. *Disasters & Emergencies.* [Online] Available at: https://apps.who.int/disasters/repo/7656.pdf

Phelps, R. D. S. G. J., 2005. *REP BR 478 Mound Filter systems for the treatment of domestic wastewater,* Bracknell, UK: BRE.

Picken, D. J., 2004. *"De Montfort" medical waste incinerators.* [Online] Available at: https://mw-incinerator.info/en/101_welcome.html

Reed, R. & de Pooter, L., 2018. *Emergency Faecal Sludge Treatment Using Hydrated Lime (Calcium Hydroxide): An introduction and case study from Bangladesh,* London: Unpublished report for British Red Cross.

Robinson, D., 2010. *Urine: the ultimate 'organic' fertilizer.* [Online] Available at: https://theecologist.org/2010/sep/22/urine-ultimate-organic-fertiliser

Rogers, T. W., de los Reyes 111, F. L., Beckwith, W. J. & Borden, R. C., 2014. Power earth auger modification for waste extraction from pit latrines. *Journal of Water, Sanitation and Hygiene for Development,* Volume 04.1, pp. 72–80.

Rose, C., Parker, A., Jefferson, B. and Cartmell, E., 2015. The characterization of feces and urine: A review of the literature to inform advanced treatment technology. *Critical Reviews in Environmental Science and Technology,* Sep, 45(17), pp. 2:455(17):1872–1879.

Rynk, R., Kamp, M., Willson, G., Singley, M., Richard, T., Kolega, J., Gouin, F., Laliberty, L., Kay, D., Murphy, D., Hoitink, H. and Brinton, W., 1992. *On-Farm Composting Handbook.* s.l.: Northeast Regional Agricultural Engineering Service (NRAES).

Scottish Environment Protection Agency, 2019. *Engineering in the Water Environment Good Practice Guide: Intakes and outfalls,* s.l.: SEPA.

Sommer, M., Schmitt, M. and Clatworthy, D., 2017. *A toolkit for intergrating menstrual hygiene management (MHM) into humanitarian response.* New York: Columbia University, Mailman School of Public Health and International Rescue Committee.

Sperling, M. v., n.d. *Main Waste Stabilization Ponds Systems.* [Art].

State of Maryland Dept. of the Environment, 2016 February. *Design and construction manual for sand mound systems,* s.l.: State of Maryland Department of the Environment.

Sunbelt rentals Ltd., n.d. *3" Diaphragm Pump Diesel.* [Online] Available at: https://www.sunbeltrentals.co.uk/plant/pumps-water-drainage-pipework/diaphragm-pumps/3-diaphragm-pump-diesel-567/[Accessed 2022].

Taylor, C. Y. J. J. D. D. A., 1997. Sand filters. *Pipeline,* Summer.8(3).

The Camping and Caravanning Club, n.d. *Provision of Chemical Toilet Facilities.* [Online] Available at: campingandcaravanningclub.co.uk

The Institution of Water Pollution Control, 1980. *Unit Process: Primary Sedimentation.* Maidstone: Institute of Water Pollution Control.

Treatment Plant Operator, n.d. *Back To Basics: What Is The Settleability Test?.* [Online] Available at: https://www.tpomag.com/editorial/2015/05/back_to_basics_what_is_the_settleability_test[Accessed 29th Sept 2022].

UK Government, n.d. *COVID-19: infection prevention and control (IPC).* [Online] Available at: https://www.gov.uk/government/publications/wuhan-novel-coronavirus-infection-prevention-and-control

UNDP Crisis Prevention and Recovery, n.d. *Guidance Note on Debris Management.* [Online] Available at: file:///C:/Users/bobre/Downloads/SignatureProductGuidanceNoteDebrisManagement11012013v1.pdf

UN-HABITAT, 2010. *Solid waste management in the world's cities - water and sanitation in the world's cities,* s.l.: UN-Habitat.

UNHCR, 2015. *D-401/2015a - Emergency Trench Latrine Wood + Plastic - Tools and guidance for refugee settings.* [Online] Available at: http://wash.unhcr.org/download/emergency-trench-latrine-design-wood-and-plastic/

UNHCR, 2015b. *OG-802/2015 Operational Guidelines for Staff: Health and Safety in Refugee WASH Programmes,* s.l.: UNHCR.

UNHCR, n.d. *Water, Sanitation and Hygiene.* [Online] Available at: https://www.unhcr.org/water-sanitation-and-hygiene.html

UNICEF, 2017. *Guidelines for Water, Sanitation and Hygiene in Cholera Treatment Centres. Somalia,* s.l.: UNICEF.

United Nations Environment Programme (UNEP), 2002. *A directory of environmentally sound technologies for the integrated management of solid, liquid and hazardous waste for Small Island Developing States (SIDS) in the Pacific,* s.l.: s.n.

United States Environmental Protection Agency (EPA), 1999. *Wastewater technology factsheet: Chlorine disinfection,* s.l.: s.n.

US Department of Labor, n.d. *OSHA Technical Manual (OTM) Section V; Chapter 2: Hazard Recognition in Trenching and Shoring.* [Online] Available at: https://www.osha.gov/otm/section-5-construction-operations/chapter-2

US Environmental Protection Agency, 2000. *Decentralized Systems Technology Fact Sheet: Evaportranspiration*, Washington, D.C.: EPA.

van Lier, J. V. A. v. d. L. J. H. B., 2010. Anaerobic sewage treatment using UASB reactors: operational aspects. In: H. Fang, ed. *Environmental Anaerobic Technology*. London: Imperial College Press, pp. 59–89.

WHO and UNICEF, 2014. *Ebola virus disease: Key questions and answers concerning water, sanitation and hygiene*, s.l.: WHO and UNICEF.

WHO and UNICEF, 2020. *Water, sanitation, hygiene, and waste management for SARS-CoV-2, the virus that causes COVID-19 - Interim guidance*, s.l.: WHO and UNICEF.

WHO, 2013. *Guidelines for the safe use of wastewater, excreta and greywater (Vol. 4)*, Geneva: WHO.

WHO, n.d. *Four steps for the sound management of health-care waste in emergencies*. [Online] Available at: https://www.who.int/water_sanitation_health/medicalwaste/hcwmposters.pdf

World Health Organization, 2006. *WHO Guidelines for the safe use of wastewater, excreta and greywater Volume 4*. Geneva: s.n.

World Health Organization, 2020. *Health Cluster Guide: a Practical Handbook*. [Online] Available at: https://reliefweb.int/sites/reliefweb.int/files/resources/Health%20Cluster%20Guide.pdf

Index

www.ingramcontent.com/pod-product-compliance
Lightning Source LLC
Chambersburg PA
CBHW051243020426

42333CB00025B/3024